Ecologization and Digitalization of Agriculture

Smart Innovation Set

coordinated by
Dimitri Uzunidis

Volume 45

Ecologization and Digitalization of Agriculture

Convergence or Opposition?

Eléonore Schnebelin

iSTE

WILEY

First published 2026 in Great Britain and the United States by ISTE Ltd and John Wiley & Sons, Inc.

ISTE Ltd
27-37 St George's Road
London SW19 4EU
UK

www.iste.co.uk

John Wiley & Sons, Inc.
111 River Street
Hoboken, NJ 07030
USA

www.wiley.com

Library of Congress Control Number: 2026935278

British Library Cataloguing-in-Publication Data
A CIP record for this book is available from the British Library
ISBN 978-1-83669-045-0

The manufacturer's authorized representative according to the EU General Product Safety Regulation is Wiley-VCH GmbH, Boschstr. 12, 69469 Weinheim, Germany, e-mail: Product_Safety@wiley.com.

Contents

Introduction

> Digital agriculture strategies will help to rationalize resources (financial and human) and address holistically the ICT opportunities and challenges for the agricultural sector in a more efficient manner, while generating new revenue streams and improving the livelihoods of the rural community. (FAO and ITU 2016, p. ix)

The FAO thus sees the development of digital technology as a solution to the socioeconomic challenges of rural areas.

> Several studies have demonstrated the need to significantly increase the world's food production by 2050. However, there is limited amount of arable land, and water levels have also been receding. [...] Our goal is to enable data-driven farming. We believe that data, coupled with the farmers' knowledge and intuition about his or her farm, can help increase farm productivity, and also reduce costs. [...] we are building several unique solutions to solve these problems using low-cost sensors, drones, and vision and machine learning algorithms. (Microsoft 2024)

Microsoft thus highlights the challenges of agriculture and its commitment to technological development in the agricultural sector.

> Blending the right portions in the right quantities and distributing them at the right moment requires knowledge, experience and time. Automated feeding and other innovative equipment take care of the daily feeding of the herd, ensuring optimal efficiency, while improving animal health. (Lely 2024)

Precision, combined with automation, is also offered for animal feeding equipment, here by Lely, a Dutch company leading the livestock robotics market.

> #CoFarming is about harnessing the power of the Internet as an intermediary, connecting farmers beyond their own neighbourhood. From land neighbors to digital neighbors! A driving force for the performance and competitiveness of farms that benefit from this networking. (CoFarming 2017)

For this French startup organization, digital technology in agriculture also represents new possibilities for collaboration and cooperation.

As the above examples show, in many discourses, whether commercial, political or media-related, the development of digital technologies is presented as a solution to current agricultural challenges, promising a better future for farmers, as well as benefits for society as a whole (Bournigal et al. 2015; CoFarming 2017; Lajoie-O'Malley et al. 2020). Digital technology would improve agricultural working conditions and farm profitability, free up more time and simplify certain tasks, better manage risks, and even promote exchanges with consumers. At the societal level, digital technology promoters claim that these innovations can meet the dual challenge of feeding the planet and respecting the environment. Beyond rhetoric, digital agriculture also mobilizes resources. States, regions, international organizations, companies, and foundations are investing in the research and development of digital solutions. Thus, the European Commission is said to have invested €192 million[1] between 2014 and 2020 in research analyzing and developing digital technology in agriculture. States, all over the world, support young companies in the sector to develop by supporting them through incubators, by financing their R&D projects and by investing in their first stages of development (Wolfert et al. 2023; Klerkx and Villalobos 2024). In France, the France 2030 plan proposes a "revolution in our knowledge of nature" in agriculture based on robotics, digital technology, genetics and biocontrol, and financially backed by 2.3 billion euros. Private funding in the field is increasing exponentially (AgFunder 2021). Research is structured around an increasing number of projects in the field, such as the DigitAg convergence institute in France, the Data4Food project at the European level and the Digiscape Future Science Platform in Australia.

However, the knowledge currently available on the uses of digital technology and the ability of these technologies to deliver on their promises is limited, and the

---

1 See: https://ec.europa.eu/info/sites/default/files/food-farming-fisheries/farming/documents/factsheet-agri-digital-transformation_en.pdf.

question of the real contributions of digital technology to agriculture remains open. Critical discourses are gradually emerging and calling into question this "technoscientific solution" (Fairbairn and Guthman 2020; Ingram et al. 2022). They also remain poorly scientifically documented, or are often focused on political issues such as data control or critiquing new technological dependencies in agriculture.

Research shows that digital technology could provide more knowledge, enable better management of complexity and diversity, and improve input optimization (Walter et al. 2017). Digital technologies, particularly precision technologies, are admitted as enabling improvements in agricultural efficiency and a reduction in its environmental impacts. Digital technology would promote the increased dissemination of information, facilitate exchanges and learning, catalyze innovation and encourage cooperation (Wittman et al. 2020). It could also reduce the workload associated with certain tasks, making it easier to implement labor-intensive agroecological practices (Bonny 2017). On the other hand, other arguments position digital agriculture in opposition to agroecology. According to some studies, the gains in terms of precision, efficiency and input savings would be relatively low and could even be counteracted by rebound effects[2] (Moschitz and Stolze 2018; Visser et al. 2021). In addition, generic digital tools would be difficult to adapt to the singularities of agricultural production systems, to local particularities and would have difficulty integrating knowledge specific to each agroecosystem (Klerkx and Rose 2020). Furthermore, the intrinsic ecological impact of these technologies (resource use and energy consumption) questions their relevance for ecologization (Berkhout and Hertin 2004; Lebrun et al. 2020). Another dimension of the controversy concerns the models and trajectories that digital technology supports. The development of these technologies is sometimes described as a continuation of the trajectory based on the mechanization, simplification and homogenization of production systems (Wolf and Buttel 1996; Clapp and Ruder 2020; Klerkx and Rose 2020). Thus, digital technology would encourage productivist rather than ecological objectives (Bronson 2019). One element of this criticism focuses on farmers' autonomy. Digital technologies would create a dependency on the suppliers of these technologies, networks, data and upstream and downstream industries (Carolan 2020; Stone 2022). It can also be seen as a loss of decision-making autonomy (Jeanneaux 2018). In addition, one criticism focuses on the cost of digital technologies, which are not necessarily accessible to all farms, and which could encourage capital concentration.

2 A rebound effect occurs when efficiency gains stimulate demand, the growth of which will counterbalance, or even outweigh, the positive environmental effects linked to efficiency gain. Indeed, efficiency gain will potentially lead to lower costs, greater capacity and new demands. This is what has been observed with efficiency gains in computers, servers, but also in the transport and energy sectors (Berkhout and Hertin 2004).

In light of the various controversies mentioned, this work starts from the idea that it is mistaken to assume, by default, that ecologization and digitalization are necessarily compatible trajectories. A priori, they are neither compatible nor incompatible. They are two different trajectories, driven by different actors, with different objectives. They will however interact, whether through practices implemented on farms or through public policies, developed knowledge and agricultural organizations' projects. In order to study the link between digital technology and agricultural ecologization, to analyze the mechanisms of interaction, the possible convergences, the paradoxes and the oppositions, it is therefore necessary to approach the question by integrating public policies, knowledge, actors and uses, which are the constituent elements of the agricultural innovation system.

This book aims to contribute to this debate by moving away from rhetoric and promises to study the representations, uses and transformations of practices that the development of digital technology has brought about in the agricultural sector. It will focus more specifically on clarifying how these innovations interact with the integration of environmental issues in agriculture. This book is organized into two main parts.

The first part presents an empirical and scientific overview of the issue of digital technology in agriculture. After a historical detour on the evolutions of agriculture since the end of the Second World War, Chapter 1 focuses on describing the contemporary transformations of the agricultural sector associated with the development of digital technology: the transformations of actors, institutions and public policies and practices on farms. These transformations raise a set of controversies on the way in which these changes accompany, or do not accompany, the ecological transformation of the agricultural sector. Chapter 2, through a review of the scientific literature on the issue, highlights the different components to be taken into consideration in order to analyze these transformations in all their complexity. It thus invites us to study these phenomena in a systemic manner by integrating several scales of analysis and the internal heterogeneity of the agricultural sector, which is too often neglected. Chapter 3 then describes how we implemented this original approach to innovation.

The second part consists of an analysis of the links between digital development and ecological agriculture, based on the perceptions, practices and projects of field actors. This part shows the digitalization of the agricultural sector in progress in its different dimensions: reflections of agricultural organizations, implementation of public policies, collaborations between digital actors and agricultural actors, uses of digital technology on farms and experiments in cooperatives. It thus shows the positions of the actors in these transformations and their practices, showing innovation not as an exogenous process that disrupts the agricultural sector, but rather as a complex, multidimensional process, which shapes and is shaped by actors. This part is organized into four chapters. The first (Chapter 4) focuses on the perceptions and implications of the organizations of the agricultural innovation system. It shows that depending on the type of ecological agriculture promoted by actors, their expectations their perceptions of risks as well as their strategies regarding digital technology differ. This diversity is little perceived by digital actors, suggesting a risk that digital technology is developed solely according to the expectations and needs of the dominant agricultural paradigm. Chapter 5 focuses on the uses and perceptions of farmers. It shows the diversity of digital use profiles. The most significant uses of digital technology are mainly carried out within more industrialized farms. The ecological practices associated with digital technology are mostly limited to justification or optimization practices rather than to deeper ecological transformation practices. However, digital tools offer knowledge access and exchange capabilities that can provide an interesting complement to more traditional knowledge exchange spaces such as farmers' associations. Chapter 6 focuses on cooperatives, which act as intermediaries between production and downstream sectors, between producers and other organizations in the agricultural innovation system. These organizations attempt to articulate a set of standards: environmental, regulatory, commercial, etc. Digital technology is sometimes used to manage the growing complexity of sectors that are diversifying and segmenting, but is mainly guided by downstream requirements rather than by ecological issues.

Chapter 7 highlights and interacts with all the previous chapters to discuss the links between digital and ecological agriculture. It shows that the development of digital technology in agriculture depends on the diversity of models and paradigms, interacting with a socioeconomic system that proposes, encourages or even imposes a technological trajectory. The agricultural digital trajectory shows several forms of opposition to a strong ecological agriculture, whether in terms of technique, objective, reasoning, temporal dynamics, but also political and social issues. Hybridizations, however, seem possible for industrialized ecological agriculture, or through a more global redesign of digital technology through rethinking its technical, economic and political models.

Acknowledgments

This work benefited from state aid managed by the National Research Agency for France 2030 and with the reference ANR-16-CONV-0004. I extend my sincere thanks to the Innovation Research Network (réseau de recherche sur l'innovation [RRI]) for this wonderful opportunity to publish my research in a book, as part of the Honoris Causa thesis prize.

I warmly thank Jean-Marc Touzard and Pierre Labarthe for their support in my career as a researcher, as well as all the people who participated in this work in one way or another. Many thanks also to my colleagues at the UMR Agir in Toulouse and the UMR Innovation in Montpellier. Thanks to Paul for his unfailing support.

PART 1

The Development of Digital Technology in Agriculture: An Innovation Process

1

Towards Digital Agriculture?

This book examines the interactions between two trajectories in the agricultural sector: the development of digital technology and its ecologization. But what exactly do we mean when we talk about digital technology in agriculture? And what is the ecologization of agriculture? Why focus on the interactions between these two trajectories? And how can these two trajectories interact?

Agriculture refers to all human activities that, by acting on ecosystems and the biological cycles of species, allow the production of resources, particularly food, from living organisms (plants, animals, fungi and microorganisms). The agricultural sector thus encompasses all these activities, all the actors involved and the institutions that regulate them.

This chapter first aims to clarify the nature of the changes in the agricultural sector, namely, the ecologization of agriculture, on the one hand, and the development of digital technology, on the other hand. Section 1.1 sets out the historical context of the agricultural sector that has led to its necessary ecologization. Section 1.2 will explain what digital technology means in agriculture. Section 1.3 will demonstrate that, beyond technologies, digital technology implies the transformations of actors, public policies and practices within the agricultural sector. This overview will lead to a clarification of the controversies surrounding the links between digitalization and ecologization based on existing literature. Practices related to digitalization and ecologization are implemented at the farm level. However, these individual practices are interconnected with broader changes in the agricultural sector and are embedded in a wider socioeconomic system. This chapter is based entirely on a bibliographical study, carried out using scientific literature and gray literature: institutional documents, media sources and websites.

1.1. Agricultural evolution: modernization and ecologization

In order to place the subject of this work in its historical and economic context, this section briefly describes major agricultural transformations from the second half of the 20th century to the present day.

1.1.1. *Agricultural modernization*

1.1.1.1. *In industrialized countries*

In the second half of the 20th century, agriculture in industrialized countries underwent a transformation known as "modernization". This modernization was enabled and supported by technological innovation in agriculture and throughout the entire value chain, particularly motorization, mechanization, the use of chemicals and genetic selection (Mazoyer and Roudart 2017). Genetics allowed for the development of varieties that incorporated the objectives of production intensification. These new varieties were combined with synthetic fertilizers and pesticides, technologies resulting from cross-sectoral dynamics with the chemical industry (Jas 2001). On farms, agricultural practices became mechanized with the arrival of the mechanical thresher and then the tractor, along with a range of other mechanical innovations (haymaking equipment, milking parlors, etc.). Research has notably played a role in ensuring the compatibility of these various innovations in genetics, mechanization and chemical production (Kirat 1991). Downstream transport and cooling technologies allow for the separation of production and consumption areas, thus enabling territorial specialization.

Upstream and downstream industries, then undergoing significant restructuring, strongly impacted the agricultural sector in its technical and spatial organization, notably contributing to the specialization of farms and territories. These developments have built what is now called "conventional farming" or "productivist farming," that is, capital-intensive, specialized, mechanized agriculture with simplified technical processes and significant reliance on external inputs (pesticides, fertilizers and animal feed) (Beus and Dunlap 1990).

This modernization is also the result of institutional and socioeconomic transformations: policies to support farmers, changes in trade rules, the concentration of upstream and downstream industries, credit regimes, research, subsidies to industries, etc. From the 1970s onwards, institutional transformations have accompanied globalization and the liberalization of agriculture, with increasing deregulation of international trade and the financialization of activities.

Thus, technological developments are impacted by, and in turn impact markets, agricultural policies, and organizations. All of these developments frame the development of a "techno-economic model for farm development" that integrates all these dimensions (Coulomb and Nallet 1980). Economic and sectoral developments are therefore reflected in agricultural practices, farm structures and the organization of the sector.

1.1.1.2. *In France*

In this work, we will focus particularly on the case of France for several reasons. First, it is a country where modernization has led to a productive, export-oriented (France is the world's sixth-largest exporter of agricultural and agri-food products) and technologically advanced agriculture. France has a strong national research and development capacity and is positioning itself at the forefront of digital development for agriculture. Second, it is a country where ecologization has taken shape in various ways: with the introduction of the environmental conditionality of subsidies from the European Union's common agricultural policy (CAP), with the inclusion of agroecological objectives in national policies[1], as well as with the strong development of quality and labeled agriculture, via indications of origin or organic farming (in France in 2022, 14% of French farms were engaged in organic farming and 27% engaged in productions with Protected Designation of Origin (PDO), Protected Geographical Indication (PGI) or *Label Rouge*[2]).

In France, in the aftermath of the Second World War, the State set itself a political project aimed at increasing agricultural production while freeing up labor, in a context of threatened food shortages and significant labor needs in the secondary sector (Servolin 1989; Flamant 2010). Added to this were objectives to stimulate growth, to free up purchasing power by reducing food costs or to strengthen France's trade balance through agri-food exports (Bartoli and Boulet 1989; Allaire and Boyer 1995).

In order to achieve these objectives, a variety of mechanisms and resources have been implemented, mobilizing the different dimensions of what can be conceptualized as an agricultural innovation system (research, experimentation, knowledge dissemination, but also regulations, institutions, technologies, etc.). Resources are mobilized to ensure the development and dissemination of knowledge necessary for agricultural modernization. The National Institute for Agricultural Research (INRA)[3] was created for this purpose in 1946. Technical institutes, agricultural colleges, agricultural high schools, chambers of agriculture, farmers' technical exchange groups (CETA)[4] and agricultural extension groups support the

1 Law of 13 October 2014 on the future of agriculture, food and forestry.

2 See: https://www.inao.gouv.fr/Nos-actualites/chiffres-cles-2022.

3 Institut national de recherche agronomique, in French.

4 *Centres d'études techniques agricoles*, in French.

dissemination and implementation of this knowledge (Gerbaux and Muller 1984). Political, legal and financial developments made it possible to reconcile the objective of competitiveness with the maintenance of family production structures through new rules governing access to credit, inheritance, land rights (the 1946 tenancy law) and control of the land market through land development and rural settlement companies (SAFER[5]) (Gueslin 1988; Boinon 2011). Modernization was also shaped by price support policies, up until the 1960s, which limited price volatility and secured investments. These policies relied on the construction and definition of so-called "professional" agriculture, delimited by thresholds in terms of size, labor or income to target a farming population that would benefit from agricultural policies (Laurent and Rémy 2000). They thus promoted a model of medium-sized family farms (Boinon 2011). These developments contribute to the movement of intensification and accumulation, which characterizes this agrarian period, itself inscribed in broader structural developments, linked to the development of the Fordist model of mass production and consumption, favored by an important role of the State and new institutions, but also the evolution of systems of thought and representations (Allaire 1988).

Increase in average farm size: from 19 ha in 1970 to 69 ha in 2020.

Decrease in the number of farms: 1,588,000 in 1970 compared to 389,000 in 2020.

Decline in the agricultural population: 2.8% of the French population today.

Simplification of crop rotations and reduction of intercropping.

Specialization of farms and territories.

Increased capital intensity of farms.

Increased yields: for example, from 4 to 7 t/ha for wheat.

Increase in input consumption: +60% in volume.

Increasing integration of farms into upstream and downstream structures.

Box 1.1. *The evolution of French agriculture with modernization*

5 *Sociétés d'aménagement foncier et d'établissement rural,* in French. SAFERs are organizations created in 1960 by the agricultural orientation law, which intervene in the land market through pre-emption and acquisition of land. The land is then resold to meet the objectives set by the law (establishment of farmers, expansion of small farms, etc.) (Boinon 2011).

The modernization of agriculture has therefore involved changes in technologies, knowledge, public policies, stakeholders, standards and institutions, and the farms themselves. These dimensions interact and evolve together and can be analyzed as a sectoral innovation system (Malerba 2002), contributing to the regulation and transformations of agriculture. Economic research has focused on studying these different dimensions to analyze the historical evolution of the agricultural sector (Allaire and Boyer 1995; Mazoyer and Roudart 2017) and its more recent transformations (Berriet-Solliec 1997; Labarthe 2006).

The so-called "productivist"[6] model entered a structural crisis in the 1980s and 1990s, characterized by frequent overproduction and growing criticism regarding health, food safety and the environment (Ansaloni and Smith 2021). This crisis led to reorientations of agricultural policies, changes in regulations and also in the organization of supply chains and production models (Touzard and Labarthe 2017). This change is sometimes defined as the implementation of an "economy of quality" in the agricultural sector (Allaire 2002). Quality encompasses several dimensions, including environmental quality. Environmental critiques of this productivist agricultural model were notably formulated by the agronomist René Dumont in France (Séjeau 2004). The environmental dimension will gradually gain importance in debates, societal expectations, research, policies and practices, in a process that we call the ecologization of agriculture.

1.1.2. *Integrating environmental concerns*

The environment, and more specifically the agroecosystem, is seen as a support for productive activities by theorists of agricultural modernization. The dominant epistemological paradigm upon which 20th-century agronomy developed reduces agroecosystems to "chemical systems whose inputs and outputs can be controlled" (Cohen 2017, p. 54). Living organisms are seen as a dimension to be controlled, mastered and standardized in order to implement homogeneous agriculture while remaining in control of circumstances (Bonneuil and Hochereau 2008). Agriculture is seen as one sector among many, which needs to be modernized in order to integrate into the global economy and markets (Bartoli and Boulet 1989). Modernization is associated with industrialization, which brings forms of

6 Productivist farming is thus characterized by "a set of technical processes, organizational arrangements and beliefs engaging actors around the requirement of increased and intensified production". It is part of a productivist mindset, defined as "the belief in the benefits of scientific and technical progress; the certainty that agriculture is a matter for professionals and specialists; and the belief that its primary mission is to feed the world's population while generating profits for actors in the sector" (Fouilleux and Goulet 2012, p. 32).

standardization and leads to adapting environments to crops rather than the other way around (Caplat 2014).

But the lack of consideration for the environment does not mean its disappearance. The publication of Rachel Carson's book *Silent Spring* (1962), which highlighted the effects of pesticides on biodiversity, brought the environmental dimension of agriculture to the forefront. The environment became a matter of societal concern and then a matter of national responsibility in the 1970s with the creation of the Ministry of the Environment in France, before becoming a subject of international negotiation following the 1992 Rio Conference (Larrère and Vermersch 2000). Agriculture is then often presented as contributing not only to the loss of biodiversity, but also to water, soil and air pollution, climate change and soil degradation. To address these challenges, there has been a growing debate on the ecologization of agriculture. This can be defined as the increasing integration of environmental issues into agricultural practices, knowledge and policies (Lamine 2011; Lucas 2021).

In French agricultural policy, this environmental shift began with the 1999 reform of the CAP[7] (Agenda 2000[8]). It took shape along in two dimensions. On the one hand, the principle of cross-compliance, strengthened during the 2006 and 2014 reforms, makes receiving subsidies from the first pillar of the CAP (income support) conditional upon compliance with minimum environmental standards (crop diversification, maintenance of ecological focus areas, etc.) (Avril et al. 2021). On the other hand, the second pillar aims to "address the specific needs and problems of rural areas", including resource management, ecosystem protection and combating climate change. Despite this shift, subsidies would be more favorable to farms with less environmentally sustainable practices, particularly due to the first pillar, which is insufficiently counterbalanced by the second (Kirsch et al. 2017). The CAP thus structures tensions between the productive and environmental challenges of agriculture. Ecologization public policies approaches are essentially based on a framework of ecological modernization, which articulates a neoliberal vision and environmental critique (Mormont 2013; Mesnel 2018).

At the level of agricultural practices, this environmental shift is multifaceted. Indeed, several agricultural models are proposed and implemented to address environmental challenges. Broadly speaking, two forms of agricultural ecologization

7 *Politique agricole commune* (PAC), in French.

8 This reform of the CAP comes in a context of crisis characterized by growing production surpluses, increasing budgetary costs, significant disparities between farms, as well as environmental impacts and health crises in food (Dehousse and Vincent 1998).

can be distinguished (Duru et al. 2015). On the one hand, an optimization model proposes improving the efficiency of the conventional production system in order to reduce its negative externalities on the environment. On the other hand, agroecology proposes applying ecological principles to agroecosystems to stimulate natural processes (Gliessman 1998). This implies a more comprehensive redesign of agricultural and agri-food systems to integrate the environment not merely as a context and support for agricultural production, but as an integral component. The first approach is the solution that seems to be favored by those involved in the paradigm of conventional farming. The second pathway is embodied in France, notably by organic farming. The latter is considered one of the prototypes of agroecology, which has become institutionalized (Bellon et al. 2011). Beyond these two pathways, a diversity of ecologization trajectories is being implemented. These different ecologization pathways are found not only at the level of individual practices on farms but also at the sector level. Indeed, organizations and institutions guide and frame these types of ecologization, whether in the production of knowledge, agricultural advice, political representation or the organization of supply chains and markets.

NOTE.– The ecologization of agriculture can be defined as the increasing integration of environmental issues into practices, knowledge and policies.

Several forms of ecologization exist, ranging from optimizing the conventional production system to reduce its negative environmental externalities to a more comprehensive redesign of agri-food systems via the application of ecological principles to agro-ecosystems.

The ecologization of agriculture is one of the challenges facing the agricultural sector, but it is far from being the only one. Agriculture also faces other economic and social challenges, such as those of remuneration and quality of work, farm succession, integration into peri-urban areas, and landscape maintenance, not to mention the major challenges of producing quality food for the entire population and contributing to climate change mitigation (Touzard 2018). Today, digital technology is revolutionizing the agricultural sector, both in terms of individual practices and collective organizations. At a time when agriculture is facing economic, social and environmental challenges, when numerous scientific studies affirm the need to change the agricultural production model, and when societal pressure is also driving this change, it seems important to examine this development of digital technology in light of the desired and ongoing changes in agriculture. What role can digital technologies play in contemporary transformations of the agricultural sector and how can we analyze their impact on different agricultural production models?

1.2. Digital technology in agriculture

This section aims to clarify the concept of digital technology and to specify it in relation to our field of study: agriculture. First, we will outline the multiple dimensions that underlie the concept of digital technology. Then, we will detail what digital technology empirically represents in the agricultural sector.

1.2.1. *Characteristics of digital technology*

While digital technology is omnipresent in discussions, it is rarely defined. This section aims to clarify the various dimensions encompassed by the concept of digital technology, from its technical aspects to its economic dimension.

The term *numérique* (digital) in French comes from the Latin *word numerus*, meaning number. It refers to a technique for encoding information, implying a discretization of information and contrasting with what is analog, that is, proportional and continuous. An analog phenomenon cannot therefore be perfectly copied digitally, as some of its properties cannot be digitized (Vadén 2004). This encoding, as a representation of information, offers possibilities for the production and reproduction, exchange, storage and processing of information, independently of the physical object. One of the main characteristics of digital objects is the ability to change the code without changing the physical object (Ceruzzi 2012; Dufva and Dufva 2018). Digital encoding thus allows the use of encoded information in a variety of tools. This information can then be managed and used remotely from the physical object.

Digital technology is therefore characterized by the codification of information. Promoting the circulation of information, this codification has given rise to another characteristic: the networking of information, particularly since the invention of the internet, the World Wide Web in the 1990s, and the Internet of Things (IoT), a field that has been growing since the 2010s (Srnicek 2018).

Digital technology exhibits significant physical and functional heterogeneity. Digital tools employ various techniques, often discussed in the field. Big data can be defined as a set of information characterized by such volume, velocity and diversity that it requires specific technologies and analytical methods to be transformed into value (Wolfert et al. 2017). The boundary is blurred, particularly with the notion of "big", which varies according to technological evolution (Carolan 2017b). The IoT refers to a network of interconnected objects based on standardized communication protocols (Verdouw 2016). The "cloud" refers to storage and computing services provided by remote computer servers. "Artificial intelligence" refers to algorithmic learning processes that enable the resolution of problems based on a given set of information. "Blockchain" is a method of storing and transferring data, without any intermediary being necessary to guarantee their authenticity and integrity (Acta 2018).

These different techniques are combined with each other and associated with hardware and knowledge to create digital technologies, which will be built, technically and socially, with the aim of fulfilling certain functions.

Digital technology is therefore first and foremost a technical system that takes shape in hardware devices. These devices have a terminal component, the one with which we interact directly, called the "terminal" or "interface": computers, phones, etc. These hardware devices also have a support component, which can be remote, particularly for the analysis, storage, or transfer of data: data centers, but also sensors, cables and antennas. The use of digital technology is sometimes described as dematerialization or virtualization. The use of these terms renders the hardware component invisible, especially the support component, which is nevertheless intrinsic to digital technology. But digital technology does indeed rely on hardware operation, and therefore uses energy and resources (Pitron 2017). On the other hand, the terminal hardware component is highly valued, as shown by the importance of design and fashion in these fields, or the development of luxury devices.

It is also important to add the human component, that is, the "human element that digital technologies contribute to putting to work" (Casilli 2019, p. 34). This human component intervenes at various stages, from design to final use, including the work of learning artificial intelligence, manufacturing and logistics. Academic research, starting in the 2000s, highlighted the emergence of new forms of work linked to digital technology, grouped under the concept of "digital labor" (Terranova 2000; Scholz 2012). This digital labor can be defined as the interactions with digital objects that are "forms of activity that can be considered work because they produce value, are subject to some kind of contractual framework and are subject to performance metrics" (Casilli 2019, p. 5).

In this book, we focus on the human element at the final stage, in other words, on uses. Use requires adopting the technology, but it also generally presupposes having appropriate infrastructure, inputs, peripherals, content, appropriable services, capacity for use, funding, learning and a governance structure (Warren 2007). Effective use can be defined as "the capacity and possibility of successfully integrating information and communication technologies (ICTs) in the achievement of objectives defined by users themselves or in collaboration" (Gurstein 2007, p. 9). The use of a technology is therefore associated with the construction of knowledge, adjustments and interactions with the context of use (Higgins et al. 2017). The use of a digital technology will thus change the nature, temporality and spatiality of the tasks performed.

> "Digital labor" encompasses several categories of work. First, there is "on-demand" work, performed by platform workers. Second, there are "micro-tasks", carried out by "clickworkers", particularly tasks supposedly performed by algorithms, but which are more efficiently done by humans (e.g. facial recognition). The Amazon Mechanical Turk (MTurk) platform is the best known in this area, offering workers worldwide the opportunity to perform tasks such as answering surveys or identifying images in exchange for (micro)wages. In a different vein, "click farms" employ individuals to click on links, thereby falsifying connection numbers and thus increasing advertising or visibility revenue. Third, there is the unpaid and unconscious work performed by digital users. Indeed, the data produced by digital users are valued by the companies that analyze it. They feed into the learning of algorithms and databases. This type of work blurs the traditional boundaries between production and consumption, between work and leisure. Studies show the effort made to render this type of work invisible (Huws 1999; Gray and Suri 2019), as well as those of its workers (Gruszka and Böhm 2020). These phenomena of revised boundaries between work and consumption (Dujarier 2014), of increased flexibility and individualization of work (Srnicek 2018), and of the invisibility of workers (Daniels 1987) are not new, but the development of digital technology is changing their nature and scope. The question of the human element in digitalization is increasingly prominent in discussions: images of click farms have been published in the press, and social movements consisting of workers affected by what is called the "uberization of the economy" are emerging, with initial legal and legislative consequences. In France, for example, Deliveroo was fined €375,000 on April 19, 2022, for "undeclared work". In California, despite a referendum that enshrined the independent contractor status of Uber drivers in 2020, debates about driver status remain heated.

Box 1.2. *"Digital labor"*

Beyond its technical and material components, the term "digital" also refers to a socioeconomic phenomenon. Its use has become widespread, encompassing not only all digital technologies but also the way they are integrated into a specific organizational and economic model. Definitions of big data illustrate this duality. One definition refers to the 3 V rule, highlighting its technical specificities that differentiate it from traditional data analysis:

– variety: the data comes from multiple sources and is heterogeneous;

– volume: the amount of data used is very large;

– velocity: data flows are in constant and rapid motion.

But a second definition integrates the organizational and economic model associated with the technical characteristics:

> The concept of Big Data refers to what can be done on a large scale, but not on a smaller scale, to extract new information or create new forms of value, thereby changing markets, organizations, the relationship between citizens and governments, and much more. (Mayer-Schönberger and Cukier 2013)

In discourse, the understanding of the concept of "digital" is therefore broadening and incorporating these components. Thus, businesses, as well as public actors, speak of the digital economy, the digital republic, digital ambition, digital inclusion and even digital agriculture. The digital economy refers to "businesses whose management models increasingly rely on information technologies, data, and the internet" (Srnicek 2018, p. 10). This definition demonstrates that digital technology is not merely a sectoral transformation, but affects all sectors and plays an increasingly important role.

The increasing use of digital technology in a field is called "digitalization" and can be defined as a process of datafication of reality. This process is also political and cultural. It can be analyzed as a stage in the history of capitalism and technological development (Srnicek 2018). It is part of a capitalist trajectory based on the pursuit of cost reduction and control over labor (Casilli 2019, p. 35) and on an economic vision that favors deregulation, property rights and flexibility (Durand 2020). It is rooted in a social trajectory of rationalization and quantification of social life (Huguet 2017). This digitalization process is gradual, and Lupton (2014) defines several stages:

1) arrival of computers and automation of repetitive tasks;

2) development of electronic identification systems and computerized management;

3) interactions, data exchanges between different institutions and systems.

Thus, the concept of digital refers not only to a technical definition, but also to the material devices, functions, uses and associated political and socioeconomic dimensions. Digital technologies can be seen as cultural and material artifacts that have political implications.

These different definitions refer to what constitutes a technology more broadly. A "technology" is not simply a set of tools, materials and equipment used for a human activity. To quote Deleuze, cited by Dockès (1990, p. 35), "machines are social before they are technical". According to Dosi (1982, pp. 151–152), a technology is also "a set of knowledge, know-how, methods, procedures, and experiences". Technological change therefore implies a complex set of transformations that affect the organization of firms, production processes, institutional configurations, intersectoral dynamics, social relations and so on (Boyer 1989). Dobrov (1979) emphasizes that technology is a set of (1) physical components – hardware, (2) knowledge and skills – software, and (3) organizational equipment – orgware. He defines the latter as "the set of socioeconomic, organizational, and management measures designed to ensure the identification and effective use of a given technique and sociotechnical knowledge, as well as the potential capacity of the technological system to adapt, develop, and self-improve" (Djellal 1995). A digital technology is thus characterized by the techniques it implements, the tool, the function(s) it serves, the knowledge and skills required by the users of this technology, the organization it implies, the associated economic and organizational model, as well as the underlying social and cultural values (Dufva and Dufva 2018).

NOTE.– Effective digital use can be defined as "the ability and possibility of successfully integrating information and communication technologies (ICTs) into the achievement of objectives defined by the individual or in collaboration" (Gurstein 2007, p. 9). It is therefore associated with knowledge building, adjustments, and interactions with the context of use.

Beyond a technical and material transformation, the development of digital technology also reflects a socioeconomic transformation. A digital technology can thus be characterized by the techniques it employs, the tool, the function(s) it serves, the knowledge and skills required by its users, the organization it entails, the associated economic and organizational model, as well as the underlying social and cultural values.

1.2.2. *Digital technology as an object of research in social sciences*

Digital technology, in its many dimensions, has gradually become an object of research in the social sciences, examined through various questions. A first set of questions relates to the context (technical, economic, social, political and cultural) of its design and construction: Why and how is digital technology developing? A second group of questions concerns the mechanisms of development, dissemination and use: What tools and what uses for which users? Here again, we find the different dimensions: technical, economic, social, political and cultural. A third dimension of inquiry focuses on the effects and consequences of digital technology in society, on the forms of meso- and macroeconomic regulation that accompany it, and more broadly on the links between digital technology and forms of capitalism: What does it change and what does it not? Some studies examine the effects on the economy, but also on social, political and cultural dimensions.

The first set of questions concerns the objectives embedded in digital technologies, the representations and imaginaries associated with them, as well as the intended uses and the power structures built into these objects. Technological development tends to be "naturalized" in discourse: it is presented as the result of a natural and inevitable evolution. This naturalization is crystallized in "laws" such as Moore's law (the power and complexity of computing doubles every 18 months) and serves to generate legitimacy and credibility for technological projects. Digital technology is thus part of the "economy of technoscientific promises" (Joly 2015). However, the digitalization project is not a natural or necessary evolution, but rather a social, economic and political project. Analyses of digital technology design show that it was conceived within an ideology stemming from a hybrid of the American hippie counterculture and the free-market principles of Californian entrepreneurs (Durand 2020), but also, in the case of the internet, from the meritocratic spirit of the research world and the cooperative and reputational practices of computer scientists (Cardon 2010, p. 13). Despite an ideal centered on individual entrepreneurship, studies show that the development of digital technology has been largely funded by states (Mazzucato 2011).

This economic, political and cultural context thus led to the incorporation of characteristics of autonomy and networking in the design of early digital technologies, as well as the "deepening and radicalization of contemporary *individualism*" (Cardon 2010, p. 9): social innovation is linked to technological innovation. These characteristics are integrated into the technical architecture, for example, the internet and the microcomputer, and therefore perform and are reinforced through use (Flichy 2004). Indeed, representations of potential uses are incorporated into the sociotechnical development process of technologies. Thus, "the inventors of the internet materialized a set of values that exert a persistent effect on the form of the network, its organization, and its practices" (Cardon 2010, p. 14).

This question also arises with regard to "data", which is a social construct and involves political and economic choices (What data do we collect? What data do we look at? How do we characterize these data?), and whose analysis is not neutral. Thus, questions and debates emerge around algorithms: What transparency is there, what governance of algorithms is there? Furthermore, other work in the social sciences focuses on the construction of digital technologies: the economic and political stakes of resource mobilization, labor and the integration of environmental concerns into the design of technologies.

The second category of questions focuses on the practices implemented. Indeed, the rhetoric of the technoscientific promises regime focuses on potentialities, on possible futures (Joly 2010). In contrast, the concrete modalities of technology adoption, the practices and uses associated with models of production, learning, change or even resistance, are much less well documented. This question is interested in concrete situations of use beyond discourse. Some studies examine the adoption and diffusion of digital technologies in society, but they sometimes tend to naturalize the social phenomenon, with technological evolution often simply being seen as the natural order of things (Bertini 2009). However, studies more specifically focused on adoption and use have highlighted individual factors, as well as economic, social and geographical factors influencing adoption. Research is examining the phenomena of the digital divide, digital skills, understanding and appropriation (Warren 2007; Brandtzaeg et al. 2011; Richardson and Bissell 2019). In order to understand the appropriation of digital technology, research also focuses on the effective use of digital technologies, that is, the integration of these technologies into practices and intended objectives (Gurstein 2007). Indeed, the use of a technology is the result of the interaction between the technology and the user: the user shapes the technical object by appropriating it (Williams and Edge 1996).

The third type of question relates to the economic functioning and the results and consequences of this technological implementation. These can occur at different scales, at different time intervals and for different dimensions. Analyzing these effects may lead us to focus on changes, but this should not, however, obscure what remains unchanged. Indeed, "technology stems from a force of conservation and perpetuation of the Same, at work in reality" (Bertini 2009, p. 19). The effects of digital development on society are a major societal issue, one that raises debates and controversies, and carries with it a multitude of hopes and fears. This question of the consequences of digital technology, and more broadly of technologies in general, is sometimes sidestepped in order to avoid the risk of technological determinism, but also the risk of slowing down technological progress and economic development by adopting a critical perspective or the precautionary principle (Bronner and Géhin 2010). However, technology and society interact and impact each other (Williams and Edge 1996).

The question of the impacts of digital technology is highly complex, as technological change always interacts with other societal developments. Some research on the consequences of digital technology examines its impacts on economic dynamics, or more broadly, on capitalism. The effects on productivity and growth are debated: the Solow paradox highlighted the fact that the development of digital technology in the 1970s had no effect on productivity. This lack of effect on productivity and growth is debated and could be linked to the measurement of productivity (Aghion and Antonin 2017). Regarding market structure, the development of the digital economy leads to monopolization and rent-seeking processes rather than investment (Durand 2020). It also reinforces trends toward increased outsourcing and just-in-time, personalized production (Srnicek 2018). On the other hand, digital technology is a source of commercial opportunities in many areas (Bai et al. 2020).

Several studies examine the consequences for work, highlighting the use of digital technology as a facilitating tool and a factor in improving working conditions and autonomy (Flichy 2004; Bai et al. 2020). On the other hand, digital technology can be used as a tool for control and subordination (Lupton 2014; Durand 2020), as a factor in the deskilling of certain tasks, job insecurity and the fragmentation of professional structures (Casilli 2019; Rotz et al. 2019b). The development of digital technology also has geographical impacts, with phenomena of spatial polarization of economic activities and socioeconomic divisions among populations (Warren 2007; Durand 2020). However, it is also seen as a factor in opening up and revitalizing rural areas, developing new services, disseminating knowledge and creating value in rural territories (Réseau Rural Français 2018). Digital technology is also changing the ways knowledge is constructed in science and business, leading to an epistemological shift in science and favoring certain rationalities over others (Kitchin 2014; Madsen et al. 2016). Data-driven scientific analysis, which "makes the data speak", is thus of increasing importance. This would amplify path dependency phenomena, as digital technology relies heavily on the analysis of past data. Finally, the effects of digital technology on the environment are also the subject of research, whether in relation to the environmental impacts of its direct use, the impacts of the changes it brings about on production processes, or economic and societal dynamics. The controversies surrounding these impacts will be detailed in section 1.4.

These three questions are interdependent, as the consequences will result both from how the technologies were designed and how they are actually used. Similarly, the design and uses of technologies will be linked to the expected consequences. These questions can be explored and studied at different scales and across different territories. Indeed, the dynamics will not be the same from one country to another, as digital development will depend on infrastructure, markets, resources and so on. Digital technology can be studied at the scale of globalized capitalism in order to understand the overall economic dynamics associated with its development (Srnicek 2018; Durand 2020). On the other hand, some studies focus on the individual level of practices and perceptions (Brandtzaeg et al. 2011). Studies have been conducted to address these questions at the sectoral level, given that the functions and impacts of digital technology vary across sectors: for example, in the industrial sector (Frank et al. 2019), in the health sector (Lupton 2014), in urban planning (Marquet 2019) or even in the agricultural sector.

1.2.3. *Digital technology in agriculture*

Digital technology refers to both technical objects and economic dynamics. The development and integration of digital technologies will interact with and impact production models and practices, and will therefore vary depending on the sector. What about agriculture? This section will specify what digital technology empirically represents in the agricultural sector. This book focuses more specifically on the development of digital technology at the farm level. However, farming is interdependent with broader sectoral developments, hence the need for a multi-scale perspective.

1.2.3.1. *Digital technology on farms: a brief history*

In France, the arrival of digital technology on farms can be traced back to the 1980s and 1990s, first with the development of Minitel and then of the computer (Pillaud 2015). Minitel was the first digital platform for a range of applications, providing access to information on weather, agricultural prices and technical and legal advice. However, it was primarily for accounting and management, and later for administrative declarations, that the first digital technologies took off (Mazaud 2017). Farmers' access to and use of computers was not solely driven by individual adoption, but also, and perhaps more importantly, by the evolution of a professional community. Thus, the majority of farmers in the 1990s used computers through intermediaries such as management centers or via collective investment within groups of farmers (Bages 1992). The development of these uses is thus linked to the development of knowledge and skills, fostered by the forms of social interaction associated with them. These technologies contribute to the "professionalization" of agriculture and reinforce the divide within the farming community, targeting farms operating according to a business model. Gradually, the computerization of farm administration and financial management has become widespread. The regulation of agricultural activities is becoming more complex and increasingly linked to the use of ICTs (Weller 2006; Oui 2021b). For example, health declarations for livestock or customs declarations for wine production are computerized processes. The ecologization of agricultural regulations contributes to this dynamic, with, for example, the mandatory registration of plant health treatments. The CAP also plays a central role in the computerization of agricultural data, with the development of online CAP declarations and the digitization of the agricultural landscape (Mesnel 2017; Magnin 2019). Furthermore, the use of the internet and social media is becoming increasingly widespread in agriculture, as in the rest of society. The Agrinautes study estimates that in 2020, 86% of farmers used the internet daily and 68% consulted social media for professional purposes (Terre-Net 2021).

In the 1980s, in the United States, new technological opportunities were developing. In particular, the authorization for civilian use of certain military technologies, such as global navigation satellite systems (GNSS), better known by their American name, GPS (global positioning system), created new technological potential (Lowenberg-DeBoer and Erickson 2019). GNSS technology thus enables the design of yield monitors, which produce maps showing yield heterogeneity within plots, the first models of which were marketed as early as 1992. Machine guidance, which allows for easier, semi-automated, or fully automated operation of agricultural machinery, appeared a few years later, first in Australia and then in the United States. This was followed by section control technologies, a device that automatically closes certain sections of the sprayer, seeder or fertilizer spreader to prevent overlaps in input application.

With the development of diverse sensors and new data sources, data analysis and modeling have flourished. Observation and experimentation infrastructures have evolved towards the massive production of data, the analysis of which, since the 1990s, has centered around specialists in applied mathematics and computer science (Oui 2021b). Initially intended for the academic sphere, big data analysis has since been integrated into tools and consulting services established in partnership with companies and aimed at professionals in the agricultural sector. Thus, new decision support tools (DSTs) are used, which provide information or recommendations on agricultural practices. For example, variable rate application technologies combine guidance, sensors (via satellites or drones) and data analysis to create maps in order to adjust input application (fertilization, but also seeds or pesticides) according to the spatial characteristics of areas within fields.

These different technologies are often grouped under the term "precision agriculture". Precision agriculture, which originally involved taking into account intra-field heterogeneity, has existed since the 1920s. But precision agriculture, as it is defined today, integrates the use of digital technologies to account for this spatial and temporal heterogeneity (Sulecki 2018). It has been developing in France since the late 1990s (Oui 2021b). This development is notably the result of investments by companies in the space sector (Airbus, Geosys, etc.) that have partnered with organizations in the agricultural sector to develop new markets (Labarthe 2009). Today, GNSS guidance technology is reportedly used on half of French farms[9]. Approximately 30% of farmers in France are equipped with yield monitors, and 6% use them. Variable rate application is estimated to affect approximately 9% of arable land in France. Connected weather stations and irrigation technologies are also

9 The data on usage estimates in France indicated in this paragraph come from the observatory of digital agriculture uses (AgroTIC Observatory 2017).

widespread on farms. However, robots, while widely used in dairy farming, remain very rare in crop production.

Digital technology has taken on a whole new dimension in agriculture in recent years, driven by a diversification of tools, leading some authors to predict a new agricultural revolution. The development of new technologies such as the IoT and artificial intelligence could even lead to data-driven agricultural systems. The terms digital agriculture, connected agriculture, smart agriculture, Agriculture 4.0, augmented agriculture and digital agriculture are then used to describe agricultural systems that make extensive use of digital technologies.

NOTE.– Digital technologies in agriculture are very diverse:

– computer tools for administrative, financial, regulatory, logistical management, etc.;

– digital technologies for information exchange;

– digital decision support tools;

– precision agriculture technologies;

– connected irrigation technologies;

– robotics, etc.

1.2.3.2. *Digitalization of agri-food systems*

Farming is also a component of the global agri-food system. Farm activities are interdependent with upstream and downstream activities within this system and are embedded in a broader socioeconomic context. This book will specifically highlight the role played by upstream and downstream activities, and the socioeconomic context of digital technology within agricultural operations themselves.

Digital developments in farming must therefore be understood within the broader context of the agri-food sector, where new technologies are emerging at every stage of the value chain. A concise overview of the digitalization of this sector was compiled by Prause (2020) based on an analysis of existing digital services. The article demonstrates the presence of digital services throughout the value chain, including sourcing (e.g. digital credit services), farming, marketing of primary products (digital marketplaces), processing, packaging (connected packaging), transportation (digital freight management), storage (automated warehouses), sales (e-commerce), consumption and value chain organization.

These food value chains are integrated into broader agri-food systems, which include:

– research and development: for example, digital technologies using satellite data, high-throughput phenotyping, the development of models based on data, or research based on data collected by civil society;

– education, including for example massive open online courses (MOOCs);

– regulatory modalities and public policies, which themselves also make use of digital technologies, for example, to carry out declarations, monitoring, farmer information or the instrumentation of measures for the ecologization of agricultural policies.

Digital technology in agriculture thus encompasses a diversity of technologies and activities at all stages of agri-food systems. These technologies can have different scales of application: a plant or an animal, a plot of land, a herd, a farm, or even a territory, a sector or a market.

In this research, we focus on digital technology at one of these scales: the farm. But this farm is embedded in an agri-food system, that is to say, a "socio-technical system that can be defined as encompassing farmers, advisors, research, upstream and downstream actors in the supply chains, public policies and regulatory bodies, consumers and civil society" (Lamine 2012, p. 139), and is more broadly integrated into a socioeconomic system. Therefore, agricultural operations must be considered within their socioeconomic environment, which itself is evolving with digitalization.

1.2.3.3. *Typologies of digital technologies*

The digital technologies used in agriculture are numerous and based on a variety of techniques. These can be distinguished as follows (Van Es and Woodard 2017):

– data collection techniques: geolocation, remote sensing, various sensors (fixed, onboard, on drone, by aerial system or satellite) which will capture data on the weather, the state of a crop (maturity, sap flow, etc.) or of an animal (heat, growth, etc.), on the soil, a resource or even a machine;

– data processing or data transmission techniques: decision support tools, radio-frequency identification, sensor networks and geographic information systems;

– action techniques: robots for livestock farming (milking, feeding, hygiene) or for crops (spraying, weeding, harvesting, etc.), drones, automated systems, automatic guidance, GPS section control, variable rate technology (which allows the application of seeds or other inputs to be adapted according to plot information), etc.;

– ICTs: on-board computers for agricultural machinery, websites (weather, markets, training, media, marketing, etc.), various software and applications (management, image recognition, etc.), as well as digital platforms.

Often, these techniques are combined to create a technology that fulfills a specific function. Thus, several types of technologies are proposed in the literature concerning digital technology in agriculture.

1.2.3.3.1. Existing typologies

A frequently cited typology relates to precision agriculture technologies. These are categorized as recording, guidance and execution or reaction technologies (Balafoutis et al. 2017). Recording technologies include those that collect information, such as sensors or satellite imagery used to create maps. Guidance technologies encompass all technologies that assist in guiding agricultural machinery. Reaction technologies use data from recording technologies to optimize input applications (water, fertilizers, pesticides and seeds). This classification clearly distinguishes the techniques that utilize digital technology in precision agriculture equipment. Originating from a mechanical engineering perspective, it does not consider how these techniques combine within technologies. Furthermore, it excludes other technologies as defined in the first section of this chapter.

In an economic analysis of the "digital revolution" in agriculture, Birner et al. (2021) distinguish between two types of technologies: embedded and non-embedded. Embedded technologies are those intrinsically linked to a specific physical object or machine. These correspond in particular to so-called precision agriculture technologies. Non-embedded technologies are software technologies that only require a generic terminal (smartphone, computer, tablet). These technologies correspond to mobile applications, management software and digital platforms. This distinction is relevant for a perspective that studies the supply of technologies. Indeed, the authors show that different actors are involved in the design of embedded and non-embedded technologies. However, this classification does not seem relevant for studying uses, which can combine both types of technologies. For example, variable rate application requires field maps (non-incorporated technology) and a fertilizer spreader adapted for variable rate application (incorporated technology). The designers of these two technological components may be different. However, the application results from the combination of these two components, and intermediaries in the agricultural sector (advisors, cooperatives, etc.) play a supporting role at the various stages of technology adoption, whether incorporated or not. Furthermore, whether incorporated or not, technologies can have very different functions.

1.2.3.3.2. Construction of a typology based on uses

Given the lack of a typology that takes into account the diversity of what digital technology encompasses in agriculture and the uses of these technologies, we propose a typology organized around the functions that these technologies fulfill. This typology distinguishes six categories, as shown in Table 1.1.

Category	Functions
Animal production	Animal feeding (automation, precision input), collection and dissemination of information on animals (health, heats, calving, etc.), milking, environmental management (bedding, buildings, silo), animal guidance, monitoring of practices and decision support.
Plant production	Soil analysis, analysis of plant condition (water, nitrogen, health), pest and disease identification, input management (phyto, water, fertilizer), soil preparation, treatment, weed control, grassland management, harvesting, transport of equipment or production, monitoring of practices, geolocation, task automation and decision support.
Management	Accounting, banking and financial services, economic analysis (calculating results/crop, etc.), inventory management, land management, environmental management, planning, official declarations, aid applications and equipment management.
Information	Information on weather, health conditions, markets, practices, legislation and online training.
Communication/collaboration	Collaborative platform: equipment rental, plot exchange, discussion networks, input purchase, service exchange, crowdfunding, equipment purchase/sale, product/co-product purchase/sale, connection with an agricultural contracting company[10] and connection with labor.
Valuation	Traceability, sale of production (financial markets, short circuits, etc.), sales management, marketing and exchanges with consumers.

Table 1.1. *Categories of digital technologies in agriculture based on their functional roles*

This typology of technologies based on functions allows us to consider digital technology in agriculture by including all dimensions of farms. These farms are the

10 *Entreprise de travail agricole* (ETA) in French. Farms that provide outsourcing services.

site of production processes, but also of economic management and decision-making processes that articulate a diversity of objectives, knowledge and constraints, in connection with the economic, institutional, social and informational systems in which they are embedded. Furthermore, this categorization is relevant for the study of uses, since it distinguishes between technologies according to the types of applications that can be made of them.

These tools and functions can be found at different scales: the plot, the operating system, the territory, the sector, etc. We therefore consider digital technology as a set of tools with a diversity of functions, and we do not consider digital agriculture as an agricultural model in itself (Leveau et al. 2019).

1.3. Transformation of the agricultural sector

The development of digital technology in the agricultural sector, which we will call the digitalization of agriculture, is leading to new uses and changes at the farm level. These transformations in farm practices depend on a number of factors: the individual characteristics of farmers (age, education, etc.) can play a role, but also the relationships they build with other actors, captured through the analysis of their ego-centered networks (Sutherland et al. 2018), or even broader sectoral developments (regulations, markets, incentives, subsidies, etc.). Thus, innovation is not here considered as a top-down process flowing from research to farmers, but as a complex process linked to a set of interactions between different types of actors, organizations, institutions and knowledge. Technological developments are therefore considered more endogenous to the sector and to the economy. Thus, even if we are interested in change at the farm level, this change is intimately linked to changes at a more global sector level. This section aims to describe the sectoral changes linked to the development of digital technology in agriculture.

1.3.1. *Changes at the actor level in the agricultural sector*

In the agricultural sector, digital technology can bring about changes in actors and in their ways of interacting. Indeed, the development of digital technology requires skills that were not present in the agricultural sector: new players are therefore being mobilized to provide these skills. This can be done through the expansion of activities of players in the digital or aerospace field (Orange or Airbus, for example), through the integration of digital skills into traditional organizations in the agricultural sector, or through the emergence of new organizations specific to "AgTech", that is to say, specific to new technologies for agriculture.

1.3.1.1. *Actors involved*

1.3.1.1.1. New actors: startups and digital actors

The development of digital technology in agriculture is bringing new players into the sector, particularly startups offering digital technologies for agriculture (Chapus et al. 2023). Startups are highlighted as drivers and fosterers of innovation in this field. They are seen as central to the technical transformations, and even to potential socioecological transformations, of the sector (Pigford et al. 2018; Piot-Lepetit 2022). They are thus supported by governments, development organizations and institutions. International investors, as well as investment funds and large corporations (Sippel and Dolinga 2023; Klerkx and Villalobos 2024), are also key players. AgriTech startups have proliferated across the globe, with examples such as Climate Corporation in the United States, AgroMatch in Chile and Naïo Technologies in France. Some introduce new products or services, while others focus on renewing existing services, such as online sales of agricultural inputs. Global investment in agricultural and food technology startups has therefore experienced strong growth in recent years (AgFunder 2021). Startups are also central to innovation public policy: Entrepreneur/Start-Up Visa in Argentina, Start-Up Chile in Chile, Innovation and Skills Plan in Canada, etc. (Audretsch et al. 2020).

But startups are not the only new actors entering the agricultural sector. The development of digital technology in agriculture is also driven by companies with existing digital skills and expertise that are expanding into new sectors. These include digital giants like Alibaba, Adobe Inc., Google (e.g. with its Mineral robotics project), and Microsoft (with FarmBeats). In France, examples include Orange Business Services, a partner of AgDataHub and Suez, working to "develop solutions that accelerate the digital and environmental transition of the agricultural sector"[11] or SFR, which offers services for implementing IoT solutions, or even Airbus, which is working on remote sensing services for fields.

1.3.1.1.2. Actors traditionally present in the agricultural and food sector

However, the development of digital technology is not solely an external dynamic within the agricultural sector. It also results from the deployment of digital strategies within traditional organizations in the sector or those already present in the sector.

11 Description of AgDataHub dated April 27, 2021, see: https://agdatahub.eu/a-la-une/agdatahub-orange-business-services-et-suez-partenaires-au-service-de-la-transition-agro-environnementale/.

This includes actors from the upstream agrochemical industry – Syngenta, Dow DuPont, Bayer Crop Science, Yara International – but also from agricultural machinery companies like John Deere and Claas. These actors are notably involved in the development of precision agriculture technologies (Oui 2021b). For example, Yara is developing the N-Sensor nitrogen modulation tool, and John Deere offers hardware that enables variable-rate application of inputs. Bayer offers modulation boards via Climate FieldView, since its acquisition from The Climate Corporation. These companies also offer a variety of other decision-support tools: for example, Bayer's Movida tool for managing downy and powdery mildew in vineyards. These companies also offer management software including Myjohndeere by the eponymous company or AgriEdge by Syngenta.

Players located downstream of agricultural activities, such as Cargill, Nestlé and Carrefour, play an important role, particularly when focusing on marketing and traceability technologies. Players that gravitate around the agricultural sector, such as insurance companies, investment funds (Demeter, CapAgro) and foundations (the Bill and Melinda Gates Foundation, for example), are also investing in digital development.

Agricultural professional organizations also participate in digital development, including agricultural cooperatives, trading companies, chambers of agriculture, technical institutes and other advisory bodies. In France, for example, chambers of agriculture offer MesParcelles, a farm management software tool, while the National Federation of Farmers' Unions (FNSEA)[12] has designed and promotes the Data Agri charter. Other examples include InVivo, the French union of agricultural cooperatives, which acquired the startup Smag in 2012, and DigiFermes, supported by agricultural technical institutes, which offer farms dedicated to experimentation with digital technologies.

1.3.1.1.3. Public actors

The development of digital technology in the agricultural sector is not solely the result of private company initiatives. Public organizations and policies support the development of digital technology in agriculture (see section 1.3.2).

In France, the Agriculture-Innovation 2025 report thus includes digital technology as an element of agricultural public policy (Bournigal et al. 2015). The Digital and Data Delegation[13] created in April 2017 within the Ministry of Agriculture aims to "support the development of digital technology, whether in the

12 Fédération nationale des syndicats d'exploitants agricoles, in French.

13 La délégation au numérique et à la donnée, in French.

functioning and tools of the administration or in the public policies for which the Ministry is responsible" (Ministère de l'Agriculture 2021). Within research, teams have been dedicated to precision machinery (at the Cemagref, now the National Research Institute of Science and Technology for Environment and Agriculture [IRSTEA][14]) or to modeling (INRA) since the 1980s, with their work gaining new momentum in the 2010s (Oui 2021b). Little by little, the issue of digital technology is making its way into public research programs (e.g. DigitAg), or public–private programs (like PIA Occitanum), and into agricultural education.

1.3.1.2. *Networks*

We have discussed the evolution of actors in the agricultural sector linked to digitalization, but these actors do not act in isolation from one another. Digital technology in agriculture evolves via numerous partnership and network dynamics.

1.3.1.2.1. Partnerships

In order to combine skills in the digital and agronomic fields, many projects are being developed in partnership between actors specializing in each of these areas, particularly via public–private partnerships, which are seen as strategic in this new techno-scientific paradigm of precision agriculture and then of digital agriculture (Oui 2021b). A historical example is the Farmstar tool offered by the Arvalis technical institute for plants[15] (Labarthe 2010). Farmstar is a crop fertilization advisory service which is based on field observations through remote sensing. The images obtained are analyzed to determine the fertilizer doses to apply to a specific area of the plot, depending on the condition of the crops in that area and the desired outcome. This service was developed through a partnership between the Arvalis and Terres Inovia technical institutes and the company Airbus using academic research from INRAE. Other types of partnerships are being implemented, particularly to combine digital and agronomic expertise, such as Microsoft's partnership with InVivo.

These partnerships are developing not only for the design of digital technologies, but also for all the innovations associated with this technological development. These innovations can be standards, procedures or charters. For example, the Agro EDI Europe association brings together 280 members: "agro-supply industries, technology providers, agricultural technical institutes, professional agricultural organizations, as well as public stakeholders". Created in 1992, this association's objective is to establish data exchange standards between all these actors. Another

14 Institut national de recherche en sciences et technologies pour l'environnement et l'agriculture, in French.

15 Arvalis is an agricultural technical institute that conducts applied research in the domain of field crops.

example is the Multipass project, a partnership between agricultural technical institutes, the Association for Agricultural Technical Coordination (ACTA)[16], INRAE, Orange and Smag, which manages the consent aspect of exchanging agricultural data and led to the creation of the company Agdatahub. The Data Agri charter, developed by the National Federation of Farmers' Unions (FNSEA[17]) in partnership with agricultural digital companies, sets out best practices for processing agricultural data. Beyond these formal alliances, meetings, exchanges and discussions are taking place among stakeholders, contributing to building a community of practice.

This interaction dynamic between actors also translates into mergers and acquisitions between companies. The most famous acquisition in this area is Monsanto's purchase of The Climate Corporation in 2013. Other examples include John Deere's acquisition of Blue River Technology and BASF's investment in ProAgrica. The Bayer and Monsanto merger is even thought to be linked to a strategy of positioning itself as a leader in the field of digital agriculture (Piot-Lepetit 2019). Agribusiness companies also directly support and finance the emergence of startups, through capital venture and incubators (Prause et al. 2020), such as Bayer Growth Ventures, Syngenta Group Ventures, Cargill Ventures or InVivo Digital Factory.

1.3.1.2.2. Building a community of practice

Beyond interactions around specific projects, identified via partnerships between stakeholders, these actors interact around the issue of digital agriculture and build networks on the subject. A number of professional events, informal exchanges and associations contribute to the creation of a shared cognitive framework (a repertoire of terms, routines, etc.), a common mission and a mutual commitment that aligns the practices of these actors regarding digital technology in agriculture, thus constituting a community of practice (Wenger 2010).

First, recent years have seen the emergence of a digital community at traditional events in the French agricultural sector, such as the International Agricultural Show (SIA)[18], the International Agricultural Machinery Show (SIMA)[19] and various agricultural fairs (Livestock Show, SITEVI [20], Innov'Agri, etc.). Digital and robotics

16 Association de coordination technique agricole, in French.

17 The National Federation of Farmers' Unions (FNSEA) is the majority professional union in the agricultural sector in France.

18 Salon international de l'agriculture, in French.

19 Salon international du machinisme agricole, in French.

20 International exhibition of equipment and know-how for the production of vines–wines, olives and fruits–vegetables.

companies gather and unite at these trade shows to gain visibility and present themselves as a community, a specific sector, for example, in the "Robotics Village" at Innov'Agri or in the "Startup Village" at SITEVI. In 2021, SIMA launched a day dedicated to digital services called "SIMA Tech Connect".

Secondly, events specifically addressing the issue of digital technology in agriculture are being organized. Trade shows dedicated to this topic have emerged in recent years. The International Forum for Agricultural Robotics (FIRA) has existed since 2016. *La Ferme Digitale* (LFDay), first held in 2017, brings together startups and various stakeholders in the sector around the theme of AgTech. Educational and research institutions also organize events dedicated to digital technology: ESA Connect at the École supérieure des agricultures d'Angers, the DigitAg days and the AgroTIC seminars, for example. A number of seminars and webinars have developed around this theme, such as the webinars organized by DigitAg, Agreenium and ACTA.

Furthermore, this community is organized around associations. In France, La Ferme numérique and CoFarming are associations that bring together startups in the agricultural sector. Robagri is an international association that unites stakeholders in agricultural robotics: startups, but also companies in agricultural machinery and electronics, research laboratories, competitiveness clusters and agricultural actors. Similarly, WineTech brings together startups that develop digital technologies in the wine industry (storage, consumption, sales, etc.). On the academic side, networks are also being structured, notably the International Society of Precision Agriculture, which organizes an annual international conference on the subject and has its own scientific journal.

Gradually, this professional community acquires recognized expertise, legitimacy and has an impact on institutions and public policies, forming what can be called an epistemic community (Haas 1992; Cohendet et al. 2014).

1.3.2. *Public policies and digital development*

Public policies for the development of digital technology in agriculture encompass measures supporting research and training on the subject, infrastructure development, as well as supply-side policies – supporting innovation – and demand-side policies – subsidies for investment in new technologies. These public policies can be driven by economic and/or agricultural policies at regional, national or supranational levels.

1.3.2.1. *Research and education*

Digital research in agriculture has existed since the 1980s, though not necessarily under that name. With the advent of computing power and new data, research developed data analysis and modeling. These techniques were then integrated into decision-support tools for farmers, offered by researchers, developed through public–private partnerships, or directly by private companies. Subsequently, research expanded into precision machinery, initially in the United States, focusing on precision nitrogen fertilization technologies. Guidance technologies were commercialized in the early 1990s in Australia and then in the United States (Lowenberg-DeBoer and Erickson 2019). In France, public research in this field developed in the 1990s at Cemagref (which became IRSTEA, then INRAE) (Oui 2021b). Research has expanded to include a variety of sensors (Lowenberg- DeBoer and Erickson 2019) and robotics.

The social sciences were slower to take an interest in this issue, with the exception of the American work of Steven Wolf in the 1990s (Wolf and Nowak 1995; Wolf and Buttel 1996; Wolf and Wood 1997). In France, social sciences on digital technology in agriculture emerged with the work of Nathalie Hostiou on precision livestock farming (Hostiou et al. 2015), the work of Danielle Galliano on ICT and the agri-food industries (Galliano and Roux 2006), and the work of Pierre Labarthe on advice and knowledge (Labarthe 2010) and Philippe Jeanneaux on farmers' decision-making processes (Jeanneaux 2018).

It was in the 2010s that these various research projects came together under the concept of digital agriculture (Martin and Schnebelin 2024). This led to the development of a body of work on digital agriculture, focusing on its technological components as well as its social components. For example, the work of Fielke et al. (2020) highlights the emergence of organizations and structures for research on this topic, as shown in Figure 1.1.

In France, this growth is characterized by the establishment of the DigitAg convergence institute[21], which brings together sixteen public and private partners and 29 research units. Its objective is to "produce necessary scientific and educational foundations for the harmonious deployment of digital agriculture in France, Europe, and the Global South[22]". This institute has substantial resources

21 A convergence institute is an approach proposed by the National Research Agency (ANR), which aims to develop large-scale multidisciplinary scientific projects that structure research and make it visible in relation to major socioeconomic issues, while producing knowledge and offering training.

22 See: https://www.hdigitag.fr/fr/qui-sommes-nous-2/.

(€9.9 million) to develop research on the subject, notably via the funding of around 100 doctoral theses in computer science, geomatics, statistics, mathematics, optics, physics, agronomy, etc., as well as in the social sciences. Beyond being a source of funding in this field, the institute also provides a structure for French research on the subject, a network of the different French teams working on these themes and also a force for international communication.

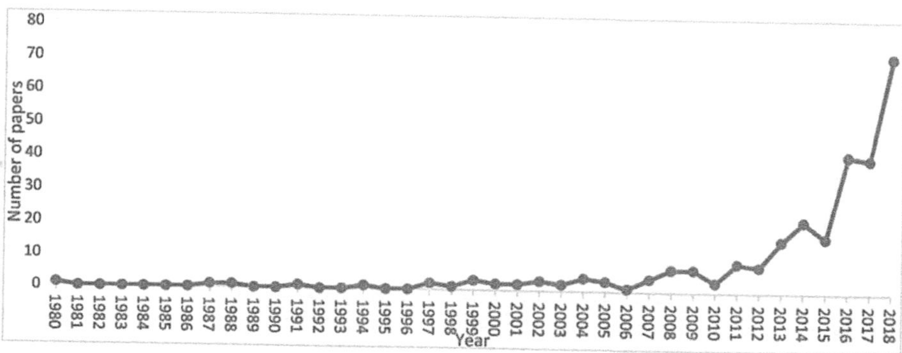

Figure 1.1. *Articles on Scopus that address digital agriculture and governance issues (from Fielke et al. (2020)). A search similar to that proposed in the article shows that this growth continues with 73 articles in 2019, 99 in 2020, 141 in 2021, 195 in 2022 and 166 in 2023 (search conducted on December 16, 2024)*

Education has also taken this technological development into account for several years. In France, for example, the AgroTIC specialization was created in the agricultural schools of Montpellier and Bordeaux to train engineers with dual expertise in agronomy and digital technology. In Australia, the master's program in Agriculture at Charles Sturt University offers a specialization in "Digital Agriculture", while the University of Western Australia offers a 3-year program called "Agricultural Technology" focused on digital skills for agriculture. In the United States, Cornell's *College of Agriculture and Life Sciences* also offers a "Digital Agriculture" option, and UC Davis offers a bachelor's degree in agricultural technology, including a "Digital Agriculture" option.

1.3.2.2. *Innovation support policies*

Digital development within the agricultural sector mobilizes human resources, but also financial and political ones, at various levels. At the European level, research projects and digital development projects in agriculture are funded. The website of the European Innovation Partnership for Agriculture (EIP-Agri) lists research and operational projects supported by the European Commission. By entering the keywords[23] suggested by the website, which refer to digital technology in a broad sense, we thus obtain 163 projects. For example, the PigWise research project aims to develop digital tools to manage the production performance, growth and welfare of pigs, and is funded by European framework programs; the LoRa Experimental Digital Agriculture Platform project aims to create a communication platform between connected objects on an experimental farm and is funded with €130,000 by the 2014–2020 Rural Development Fund; The Smart Beekeeping Kept Simple project aims to develop good beekeeping practices based in particular on sensors, funded by Horizon 2020; as does the Internet of Food and Farm 2020 project, also funded by Horizon 2020, which aims to increase the productivity and sustainability of agri-food systems via the IoT. Thus, the European Commission reportedly invested €192 million[24] between 2014 and 2020 in research for and on the development of digital technology in agriculture. Public support for innovation is available at the various stages of technological maturity, from exploration – for example, with the SmartAgriFood program exploring the potential of digital technology in agri-food value chains and funded by the European Union – to deployment, with, for example, the IoF2020 project to develop the "Internet of Things" in the agricultural sector (Wolfert et al. 2023).

Around the world, networks, funding (often involving both public and private sectors) and infrastructure are developing to support the emergence of digital agricultural businesses (Klerkx and Villalobos 2024). Many countries fund incubators to support the growth of companies in this field, such as AgLaunch in the United States, French AgriTech in France and the Wageningen University and Research Start Hub in the Netherlands. In Chile, the "Transforma Alimentos" program aims to support innovative companies in the agri-food sector.

At the national level in France, general economic measures (support for innovation, startup funding and research tax credit) provide substantial support to companies offering digital technologies and services in the agricultural sector. *Le Grand plan d'investissement 2018–2022* supports "innovation and the structuring of agricultural sectors" (€500 million) and "the upgrading of downstream agricultural activities" (€1.6 billion). Furthermore, on August 30, 2021, the French government launched French AgriTech, a future investment plan that allocates €200 million to the development of agricultural startups over 5 years.

23 Blockchain, data management, data sharing, data innovation hub, digital learning, digitization, drone, decision support system (DSS), data-driven agriculture, ICT/software, ICT, information system and platform.

24 See: https://ec.europa.eu/info/sites/default/files/food-farming-fisheries/farming/documents/factsheet-agri-digital-transformation_en.pdf.

1.3.2.3. *Policies supporting innovation in farms*

Agricultural policies can also drive the development of digital technologies. In Europe, the CAP also supports the digital development of agriculture through various means. First, the development of online CAP declarations, which became mandatory in 2016, along with various regulatory tools (online wine custom declarations, traceability of plant protection product use, etc.), constitutes a significant driver of digitalization (Mesnel 2017; Magnin 2019). Second, for the 2014–2020 CAP, the second pillar offers various digital financing tools[25]. Article 17 provides co-financing for equipment to modernize farms. This can facilitate investment in guidance or variable-rate application systems. The agri-environmental or climate measures proposed in Article 28 can be justified by the use of precision agriculture equipment. Article 35, which aims to promote cooperation, can finance the joint purchase of precision farming equipment. The measures on advisory services (Articles 14 and 15) can also be used to disseminate precision farming equipment. The role of digital technology will be strengthened in the future CAP, with Member States implementing strategies for their knowledge and innovation systems (Article 5) and explicitly focusing on digitalization as a key element of modernization, supported by the European Commission (Article 102) (Labarthe et al. 2021a). At the French level, the *Grand plan d'investissement 2018–2022* allocates over 1 billion euros to "supporting investments in agricultural businesses" (Agriculture 2021).

Various innovation support policies, targeting the private sector, the public sector or agricultural operations, converge, since innovation policy is based on relationships and partnerships between these types of actors.

1.3.3. *Digital technology uses in farms*

The development of digital technology in the agricultural sector translates into changes in terms of actors and public policies, but also into the development of new uses, that is to say, the effective use of digital technologies and their integration into agricultural operations.

Farmers' use of digital technology is relatively little known. Official agricultural statistics are beginning to include some figures on digital equipment on farms, but this information remains very imprecise and outdated.

In France, the 2010 agricultural census only provides a few variables related to digital technology: internet use, broadband internet use and the use of specialized

25 Information based on the European Parliament document, see: https://www.europarl.europa.eu/RegData/etudes/note/join/2014/529049/IPOL-AGRI_NT(2014)529049_EN.pdf.

accounting and technical management software (field monitoring, herd management, etc.). However, studies provide some information on digital technology use in French farms. *L'Observatoire des usages du numérique en agriculture* (the Observatory of Digital Uses in Agriculture), created by the Agrotic Chair and the DigitAg convergence institute, aims to demonstrate the current use of digital technology in France: the technologies used, the types of agriculture affected and the applications. Infographics are produced on the use of specific tools. Data are collected through literature reviews, interviews with companies providing these services and interviews with agricultural advisors and technicians. According to their studies, the commercialization of remote sensing services covered 905,000 ha in 2020 (compared to 1.1 million in 2016). This represents 10% of arable land and 1.2% of vineyards. Approximately 30% of farms use yield sensors. Tractor geolocation systems (guidance) are used by half of all farmers in France, with 30% using RTK correction. One in two farms has a connected weather station. In cattle farming, more than 10,000 robots were in use in 2018. In crop production, only a few hundred robots are used, mainly for weeding, with most still in the prototype stage. The smartphone is used for equipment adjustment applications (spreader, seeder, tires, etc.), surveying applications (plant sizing, etc.), for guidance (GPS), to access weather information or to enter field observations.

Since 2006, an annual survey on internet use by farmers in France, the Agrinautes survey, has been conducted for Terre-net média on a sample of nearly 1,000 farmers, whose responses are then weighted according to the region and the farm's technical and economic orientation (OTEX)[26]. These studies show that in 2020, 95% of farms had mobile network coverage, 70% of farmers owned a smartphone (45% in 2016) and 86% connected to the internet at least once a day. The study indicates that the most frequently accessed online content (i.e. by the largest number of farmers) was, in order, weather, banking services and classified ads. The results indicate that 68% of connected farmers consult social networks for professional purposes (42% in 2016), primarily Facebook (44.7%) and WhatsApp (40.9%), but also YouTube, Twitter, etc. The preferred sources of information remain printed agricultural press and distributor advisors. Furthermore, almost three-quarters of farmers make purchases online, mainly small consumables and spare parts. Regarding decision support tools, 44% use them (Ministère de l'Agriculture et de l'Alimentation 2016; Terre-Net 2021).

An Agrodistribution-ADquation survey on digital technology use in French agriculture was conducted in 2014. The data from these two surveys, although incomplete, are the primary source of information on digital technology use. It was notably used in the Renaissance numérique think tank's 2015 report, which describes farmers as "early adopters" and cites the following figures: 79% of

---

26 *Orientation technico-économique*, in French.

farmers use the internet and 76% check the weather online several times a week (Isaac and Pouyat 2015). This report appears to serve as a reference for the Ministry of Agriculture, which incorporates the data it contains into its infographic on digital technology in agriculture (agriculture.gouv 2016).

There are limited data on digital technology use in the scientific literature, particularly in France. Economic literature analyzes the factors influencing the adoption of digital technologies, but rarely examines the uses themselves. For precision agriculture technologies, this research highlights individual factors (age and education), economic factors (farm size, type of production, income and labor), socioeconomic factors (cooperative membership and consulting) and technological factors (ease of use, time required and available information on its effectiveness) (Barnes et al. 2019; Lowenberg-DeBoer and Erickson 2019). This research, which will be presented in more detail in Chapter 2, shows that these factors vary depending on the technologies involved.

Information on farmers' use of digital technologies is therefore still limited, with very little information on actual use, adoption and the consequences of these technologies on farms. National statistics currently provide little information on digital usage, and the scientific literature on the subject is also limited, both in France and internationally. The available data are incomplete, both in terms of representativeness and usage, since most data are concerned with the adoption of a technology or the use of a service in isolation. The consequences of these technologies mentioned in scientific and institutional documents are therefore essentially presuppositions and potentialities, since few studies examine the effects of these technologies in real-world conditions. The concrete changes brought about by these tools are poorly understood. Yet, digital technology is largely supported by public funds. This situation is even more paradoxical given that digitalization is the subject of controversy, particularly regarding its ability to address the environmental challenges of agriculture.

Digital technology can create knowledge or knowledge-creation tools, and it transforms knowledge transfer as well as knowledge itself. Indeed, the very nature of digital technology is the codification of information, which facilitates its transfer and storage: information, and therefore certain types of knowledge, can be easily disseminated, cataloged and so on. The codification process can change the nature of knowledge, for example, via the loss of tacit knowledge or its transformation into explicit knowledge.

Thus, the development of digital technology in agriculture encompasses a heterogeneity of technologies, but also of actors, institutions and socioeconomic transformations. "Digital agriculture" itself ultimately exists only through the technoscientific promise of this technological shift, which proposes digital technology as a solution to the environmental, but also socioeconomic, challenges facing contemporary agriculture (Martin and Schnebelin 2024). However, the development of digital technology is the subject of controversy regarding its capacity to resolve these challenges.

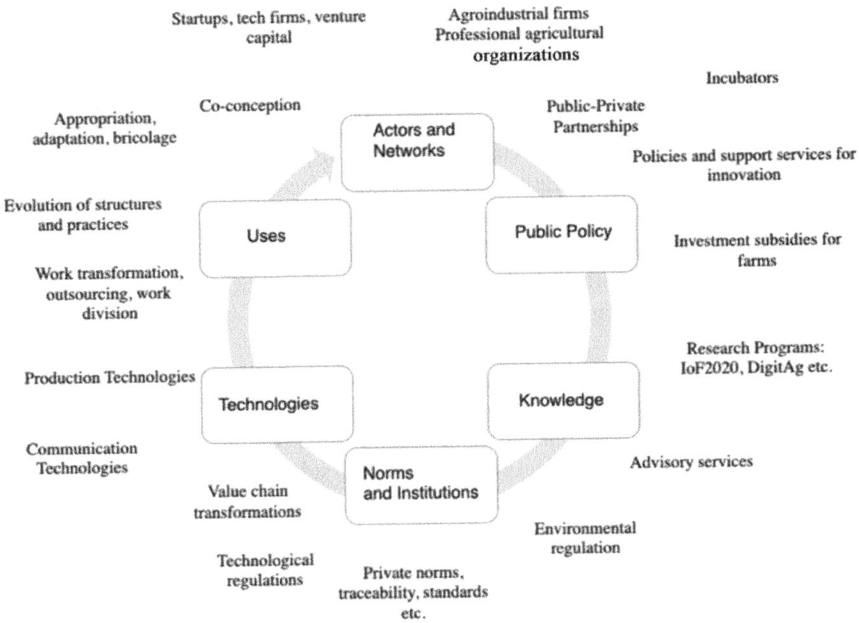

Figure 1.2. *A co-evolution of the dimensions of the agricultural innovation system with the development of digital technology*

1.4. Controversies surrounding digital technology and ecology in agriculture

The development of digital technology is being promoted by scientists, agricultural organizations and international bodies to both improve agricultural productivity and reduce its environmental impacts (Lajoie-O'Malley et al. 2020). Consequently, the deployment of digital technologies is supported and integrated into agricultural policies that aim for dual economic and environmental performance (Prause et al. 2020). However, this deployment is subject to controversy. In particular, the compatibility between the digitalization of agriculture and its transformation in order to integrate environmental issues is debated. First, the environmental impacts of digital technology are the subject of controversies not specific to the agricultural sector. Second, the impacts of digital technology are the subject of debates specific to the environmental question of agriculture, and particularly in relation to the agricultural model(s) that digital technology can support.

1.4.1. *Digital technology and ecology*

Digital technology is presented as a solution to environmental problems in numerous institutional documents, scientific reports and think tank or corporate reports (ADEME et al. 2017; Bai et al. 2020; Babinet 2021; Ministère de la transition écologique et ministère de l'économie 2021; Potier 2021). Although rarely supported by evidence, the assumptions linking digital technology and the environment are based on several elements. The main element is the idea that data created by digital technologies allow for better process management, optimization and therefore a minimization of negative impacts. Thus, digital management is presented as "smart" management. The "smart city" based on sensors and automated systems to manage transportation networks, waste management, hospitals and so on would optimize flows and improve the management of urban energy consumption by modulating city lighting according to occupancy levels or optimizing building energy consumption (Marquet 2019; Potier 2021). The "smart factory" based on the near-complete automation of all company activities and flows would manage the quantity and quality of production and limit waste (Lezoche et al. 2020).

Businesses could better manage their waste. Digital technology would optimize the transport of goods and improve the synchronization of supply and demand through information sharing. These technologies could also promote the implementation of circular economy mechanisms by facilitating information exchange.

This positive perspective, which is reflected in these institutional and professional documents, contributes to a "soft" power, that is, the capacity to inspire, persuade and mobilize through ethical or intellectual arguments (Morgan et al. 2008, pp. 4–5). This contributes to the construction of a shared vision of a desirable future where digital technologies offer new ways to address society's environmental challenges. Digital innovation thus falls within the framework of the technoscientific promise economy, which helps to lend credibility to its promoters, legitimize their innovations and mobilize diverse resources (Joly 2015).

To avoid mechanistic and simplistic links between the use of digital technologies and the environment, Berkhout and Hertin (2004) describe the types of interaction and therefore the potential types of impacts of digital technology on the environment. This leads them to identify three types of impacts: direct impacts arising from the production and use of these technologies, indirect impacts resulting from the effects of using these technologies, and structural and behavioral impacts linked to long-term socioeconomic changes associated with these technologies (see Box 1.3).

> The development of digital technology can have several types of impact on the environment:
>
> – Direct impacts are the impacts directly resulting from the production and use of these technologies. Examples include the pollution generated during the design of technologies via metal extraction, or the energy consumption for the design and use of technologies, particularly for server operation. Estimates of the ICT sector's energy consumption vary widely and are subject to debate. Some estimates indicate that this sector used approximately 8% of global electricity in 2010 and could reach around 14% by 2030 (Lange et al. 2020). These direct impacts can also be linked to the use of non-renewable resources, waste production or recycling. The authors show that all these direct impacts have a negative environmental balance.
>
> – Indirect impacts arise from the effects of using these technologies. For example, there are effects related to changes in production processes, such as the optimization of production processes via modeling or data-driven logistics and distribution management (Lange et al. 2020). Furthermore, digital technologies can enable the "dematerialization" of products, services or work: access to information, music, meetings and conferences, completion of certain tasks, etc. (Berkhout and Hertin 2004). These can replace existing goods and services, reducing environmental impacts, or add to them, creating new environmental pressures. According to the authors, these impacts are both positive (efficiency gains, virtualization and environmental management) and negative (proliferation and addition rather than substitution).
>
> – Structural and behavioral impacts are broader and more difficult to assess because they interact with a range of other factors. These impacts are linked to structural changes in society, such as the sectoral restructuring of the economy, changes in lifestyles and consumption patterns, and evolving travel patterns. These structural changes can lead to what is known as a "rebound effect". Such an effect occurs when efficiency gains stimulate demand, the growth of which will counterbalance, or even exceed, the positive environmental effects associated with the efficiency gains. Indeed, an efficiency gain will potentially lead to lower costs, greater capacity and new demands. This has been observed with efficiency gains in computers and servers, as well as in the transportation and energy sectors (Berkhout and Hertin 2004). These changes can be positive or negative on the environment and can change over time and space.

Box 1.3. *Types of digital impacts*

The work of Lange et al. (2020) focuses solely on energy impacts. The authors also categorize the impacts into several categories: direct impacts, energy efficiency, economic growth and sectoral changes. Their work shows an overall negative effect of ICTs on global energy demand.

The key takeaway from these studies is that to analyze potential or actual effects, one must not limit oneself to the simplest effects on production processes, but rather consider the overall scope and temporality of these effects. However, hypotheses linking digital technology and the environment often only consider simple causal relationships, such as increased process efficiency. These hypotheses frequently neglect the intrinsic impacts of producing these technologies and broader impacts, such as the rebound effect. These impacts are difficult to assess because they are complex, interdependent, uncertain and dependent on the envisagedcontext and scale (Berkhout and Hertin 2004). They are also linked to social, economic and political effects that can interact with environmental issues.

1.4.2. *Controversies in the agricultural sector*

Technological innovations typically carry a dual perspective in society: on the one hand, they offer transformative potential and, on the other hand, they are a source of unexpected consequences and potential new problems, including environmental ones (Tenner 1997). Thus, technological development is in constant tension between these two facets (Clapp and Ruder 2020). The resulting debate on the links between digital technology and ecology is also found in the agricultural sector. On the one hand, digital technology is seen as a source of effective solutions to address the environmental challenges of agriculture while maintaining high productivity. On the other hand, digital technology is seen as a source of risks that could reinforce, or even extend, the negative impacts of agriculture on the environment and, more broadly, encourage the agro-industrial system. This debate is reflected in academic works, institutional documents, and a number of texts and political manifestos from economic, environmental and agricultural organizations, which have taken up the subject, either to promote or criticize it. The main arguments in this debate can be organized into several identified categories: agricultural practices and labor, knowledge, autonomy, material resources and the agricultural model.

1.4.2.1. *Effects of digital technology on the environment through the lens of agricultural practices and labor*

The most prevalent argument in discussions promoting digital technology as an environmental solution is its effect on process efficiency (Walter et al. 2017). This effect corresponds to "indirect impacts", as explained in Box 1.3. In particular, many documents highlight the potential of digital technologies to optimize the use of inputs, whether it be fertilizers, pesticides or water (Bai et al. 2020; INRAE 2020; Microsoft 2021). This optimization is made possible by the ability to make data-driven, scientifically sound decisions. Digital technology is also said to promote higher yields and therefore require less land dedicated to production to "feed

10 billion people" (Microsoft 2021). It is also said to encourage non-tillage farming and thus carbon sequestration in soils (Clapp and Ruder 2020). Furthermore, the gains from process efficiency would help limit losses and waste, such as livestock effluents or nitrogen losses in the soil. Digital technology could also facilitate resource management at the territorial level, for example, by identifying material flows and facilitating collective decision-making (Bellon Maurel et al. 2022).

Digital technologies, by generating new data, offer potential for understanding and managing ecological processes, which could lead to better management of the complexity inherent in living systems and thus contribute to a more agroecological agriculture (Bellon Maurel and Huyghe 2017). They would provide the capacity to manage each agronomic decision on a case-by-case basis and to take into account the complexity of agronomic problems, for a "measured agriculture" (Grenier 2018, p. 30). At the value chain level, digital technology would enable better coordination between organizations and therefore greater efficiency (Sonka 2021). More broadly, digital technology would support the ecological, or even agroecological, transition of agriculture (Bournigal 2016).

Furthermore, these technologies would reduce labor requirements. This could facilitate the implementation of more agroecological, yet labor-intensive, tasks such as mechanical weeding (Bellon Maurel and Huyghe 2017; Lebrun et al. 2020). These arguments are echoed in the discourse of robotics companies, which propose to solve labor shortages, limit the use of precarious workers and reduce the arduousness of labor (Lenain et al. 2021).

Conversely, other scientific studies suggest that the gains from precision agriculture would be relatively small (some studies estimate them at around 10%) and could even be offset by a rebound effect that would limit the anticipated savings (Moschitz and Stolze 2018). For example, improved fertilization efficiency could encourage increased fertilization to maximize yields or the fertilization of more crops. These technologies would be designed by incorporating production-oriented objectives, integrating a vision of farming based on profit maximization and optimizing quantity at the expense of quality, thus reinforcing this mode of reasoning in farming, such as monoculture, simplified technical processes, etc. (Bronson and Knezevic 2016; Carolan 2020; Réseau Action Climat 2020).

Furthermore, if technologies are designed to optimize inputs, this does not incentivize the elimination of those inputs: it creates dependency effects between technologies and locks practices (Wolf and Buttel 1996). All of this would favor the continuation of the productivist trajectory based on increased productivity, but also on the specialization and concentration of production (Clapp and Ruder 2020; Lioutas and Charatsari 2020). Because these tools incorporate existing practices, their widespread use could lead to a homogenization of production systems (Wolf and Buttel 1996; Jakku et al. 2016). This could reduce the adaptation of agroecosystems to their local specificities, thus leading to a decrease in biodiversity. The development of digital technologies could also lead to a standardization of tasks and a deskilling of agricultural workers, contributing to their precarization (Rotz et al. 2019b).

1.4.2.2. *Effects of digital agriculture on the environment through a knowledge lens*

Agroecology requires extensive field and research knowledge, whether in relation to the functioning of agroecosystems, biodiversity or nutrition. Digital technologies could contribute to the creation of this knowledge and to new ways of knowing (Bonny 2017; Ingram and Maye 2020). Since agroecosystems are heterogeneous, the wealth of information would allow for better management of this spatial and temporal heterogeneity. This principle of managing heterogeneity is the basis of precision agriculture technologies, whose maxim is "The right dose, in the right place, at the right time" (Bordes 2017). Data from new sensors could be used for the precise understanding of ecological processes and thus for better use of ecosystem services or better integrated management of plant or animal health (Bellon Maurel and Huyghe 2017).

Digital technologies make it possible to reduce the cost of storing, disseminating and accessing information. This could promote knowledge exchange favoring the implementation of more ecological practices (Phillips et al. 2018; Leveau et al. 2019), as well as the redesign of production systems (Gkisakis and Damianakis 2020). It could promote the breaking down of silos in research (Damave 2017). Digital technology could be a tool in building commons around agroecological knowledge (Fraser 2020). The development of digital technology could also facilitate traceability, which could highlight and enable the economic valorization of products from agroecological production (Wittman et al. 2020).

Conversely, the use of digital technologies could lead to a loss of farmers' tacit and empirical knowledge, as this knowledge would be replaced by the use of technology (Barrett and Rose 2020; Clapp and Ruder 2020). This could result in a loss of agricultural know-how and sensory intelligence (Atelier paysan 2021). Furthermore, by integrating a new artifact between the producer and the environment, this could deteriorate the relationship with nature, and therefore the knowledge and consideration of the environment in decision-making (Rose et al. 2021). The quantification and standardization of practices through digital technology could rigidify practices, reduce farmers' field experimentation and thus limit the adaptation of practices to local specificities and freeze ecological knowledge (Burton and Riley 2018).

1.4.2.3. *The question of autonomy at the heart of the relationship between ecology and the digitalization of agriculture*

Autonomy is a recurring theme in debates surrounding digital technology and, more broadly, technology in agriculture (Clapp and Ruder 2020). Rarely defined, the concept of autonomy can encompass a variety of interpretations, some more individualistic and others that view it more as a form of collective capacity building (Atelier paysan 2021; Stone 2022). This diversity of interpretation leads to a heterogeneous use of the concept of autonomy, ranging from advocating for farmers' independence from external labor to the autonomy of professional communities to repair their own equipment, and including autonomy in marketing their produce or decision-making autonomy.

In debates on the environment, and particularly on agroecology, autonomy is seen as essential. On the one hand, reduced dependence on external inputs would allow farms to be more resilient to economic and ecological risks, and would promote more economical use of resources (Van der Ploeg et al. 2019). On the other hand, farm autonomy is also seen as having a certain independence from the industrial agri-food system, and would therefore allow for a broader range of actions, by integrating other objectives and values, more in line with the environment (Wolf and Nowak 1995; Atelier Paysan 2021). Autonomy would thus be the expression of a form of power for farmers vis-à-vis other economic actors (Carbonell 2016).

Digital platforms and websites could redefine how farmers communicate and market their products. Digital technology is thus highlighted as a way to access marketing channels that are independent of traditional actors (Carolan 2017a). It would offer new opportunities for cooperation among farmers, thereby strengthening their capacities (CoFarming 2017). In another area, digital technology would facilitate the implementation of new, more participatory research methods and could allow farmers to be directly involved in research projects that concern them (Bellon Maurel et al. 2022). Digital technology would be a tool for asking new questions, while leaving decision-making power with the farmer (Grenier 2018). Farmers would have greater autonomy thanks to a more comprehensive view of their activities (Damave 2017).

Finally, digital technology would facilitate direct exchanges between farmers: whether for selling, renting, or buying produce, equipment, farmland, services, etc. This would promote their own autonomy. Agricultural robotics companies also emphasize this notion of autonomy, which would be amplified through the use of robots, freeing farmers from dependence on labor, work, herbicides, etc. (Lenain et al. 2021).

On the other hand, some point out that digital technology creates technological and economic dependencies, potentially leading to forms of lock-in (Lioutas and Charatsari 2020). The use of technologies for inputs would increase dependence on those inputs (Wolf and Buttel 1996). Since a number of digital technologies originate from the GAFAM companies[27], new forms of dependence on these companies would develop, thus fostering the concentration of power. Digital technology would contribute to a privatization of the commons and an increased dependence on digital technology actors (Confédération Paysanne 2021). The high investments involved in purchasing digital technologies for a farm can lead to a loss of autonomy, by, for example, creating dependencies on sources of financing or loans, the need to make the investment profitable, and therefore the need to continue using the production methods managed by the purchased technology (Gkisakis and Damianakis 2020). These phenomena could lead to the exclusion of some farms (Bellon Maurel et al. 2022).

Information and advisory technologies also challenge farmers' decision-making autonomy (Jeanneaux 2018). Furthermore, based on a highly individualized view of advice, digital technology, as offered by companies, is incompatible with the social and collective dimension of farmer autonomy, which is based on social relationships of work and information (Stone 2022). Overall, digital technology leads to increased dependence on the agro-industrial system (Atelier Paysan 2021). Discussions surrounding autonomy encompass different approaches, both individualistic and collective. For example, American farmers obtained an exception law that allows individual modifications to the programming of their agricultural machinery, but prevents the development of epistemic communities on this topic (Carolan 2017a). The author contrasts this with the development of open-source platforms that foster user communities and collective autonomy.

1.4.2.4. *Material dimension*

Another aspect of the debate concerns the most direct impacts linked to digital hardware devices. This broadly corresponds to the direct impacts of these technologies.

These technologies can enable the "dematerialization" of certain tasks and reduce their environmental impact: for example, by limiting the use of certain resources or reducing travel. However, the design, use, and operation of these technologies require energy and resources, particularly rare metals (Atelier Paysan 2021; Confédération Paysanne 2021). The extraction of these resources is known for its negative consequences on the environment and on the populations working in this

27 GAFAM is an acronym used to refer to the dominant firms in the digital market: Google, Apple, Facebook, Amazon and Microsoft.

extraction or living near mines (Pitron 2017). The potentially widespread use of digital technologies in agriculture also raises the issue of obsolescence and waste management for these technologies.

1.4.3. *Controversies surrounding digital technology and pathways to ecologization*

The agricultural sector is characterized by debates on the ecologization pathways to follow. Indeed, as mentioned in section 1.1, there are different ways to consider the transformations of agriculture so that it limits its environmental impacts. One of the proposed paths is the optimization of the conventional model through incremental innovations and in particular technological innovations. Another proposed path is the more profound redesign of agri-food systems based, in particular, on the use of ecosystem services (Duru et al. 2015). Organic farming is among the models that offer a more comprehensive redesign, or at least a prototype of such a model (Bellon and Penvern 2014), even though there are debates about the potential conventionalization of organic farming (Darnhofer et al. 2010). Several agricultural models therefore coexist within the sector. Their coexistence depends on the power dynamics that structure the sector.

Consequently, the controversy surrounding digital technology is reflected in the debate on forms of ecologization. Part of this debate, therefore, raises the question of which agricultural model is favored by digital technology. Does digital technology create a new agricultural model, as some claim (Bournigal 2016)? What more structural transformation does it accompany? Does it apply generically to all forms of agriculture and therefore all forms of ecologization, or does it favor some over others? These questions relate more broadly to the structural impacts of digital development in the sector. However, this aspect of the debate is rarely addressed in the literature.

The often-used concept of digital agriculture might lead one to believe that there is a new, digital agricultural model, as opposed to a non-digital one. In the characterization of digital technology in agriculture presented in this chapter, digital technology encompasses a multitude of technologies, but also a multitude of functions and actors. This raises the question of the boundaries beyond which agriculture could be considered digital. The unity of digital agriculture in discourse seems to stem from an umbrella term that contributes to the construction of a technoscientific promise, but does not correspond to a homogeneous agricultural model (Martin and Schnebelin 2024). In the work presented here, digital technology is not considered as a carrier of a new agricultural model, but rather as a set of heterogeneous technologies.

Digital technology in agriculture therefore does not appear to be a new agricultural model: it is being integrated into existing agricultural models. But is it being integrated homogeneously into all models? Is it better suited to one model than another?

One aspect of this debate concerns the compatibility of these technologies with agroecological models. ICTs and precision agriculture are indeed among the main controversies related to agroecology (Migliorini et al. 2020). Other questions, very present in debates on agroecology, are concerned with the role of agrochemicals and biotechnologies, farm size, short supply chains, social justice and gender issues.

Contradictions	**Synergies**
Process production: agricultural practices, labor	
Homogenization of production systems, no adaptation to local specificities Rebound effect, locking in practices Productivist objectives – specialization	Input optimization, multi-criteria decision-making Managing complexity Facilitation of normally labor-intensive tasks
Knowledge	
Tacit and empirical knowledge loss Deteriorates relationship with nature Reduced experimentation, rigidity	More information, more knowledge: improves understanding, differentiated management Reduced cost of access to information
Autonomy	
Dependence on large agricultural supply companies, on GAFA Loss of decision-making autonomy	Better communication with consumers Participatory research Sharing, cooperation
Paradigm, agricultural model	
Promotes industrial model, reinforcement effect Elimination of alternatives	Generic technologies, for all types of models, also meet the needs of agroecology
Material setup	
Impacts of digital technology construction, energy cost	Dematerialization
References	
(Wolf and Buttel 1996; Carolan 2017b; Bronson 2018; Rotz et al. 2019a)	(Pillaud 2015; Bellon-Maurel et al. 2017; Van Es and Woodar 2017; Walter et al. 2017)

Table 1.2. *Main arguments in the controversy on the links between digital technology and ecological agriculture*

Innovations of all kinds, including technological innovations, have played a significant role in the past evolution of agricultural production systems and continue to play an important role in the trajectory of agricultural models. It is therefore necessary to examine which technologies are acceptable and beneficial for different agricultural models. This involves economic, social, environmental, organizational, cultural and political questions regarding these innovations. Here, we develop an economic approach that questions their compatibility with production models.

Agroecology definitions do not in themselves exclude the use of digital technologies. This depends on the type of technology and its effects. There are few studies to date on digital technology and agroecology. Various effects can be discussed: the impacts of these tools on production, social impacts, costs and benefits, and mechanisms of economic dependence. Technologies could promote productivity, input efficiency, concentration and specialization (Wolf and Buttel 1996; Carolan 2020). Thus, these technologies would not promote agroecological models, but rather the continuation of the conventional agricultural model (Bronson and Knezevic 2016; Lioutas and Charatsari 2020). On the other hand, technologies could provide localized solutions for optimization and consideration of heterogeneity and the mechanisms of living organisms, which could help the implementation of agro-ecological practices. Thus, the question of future agricultural practices favored by digital development remains open.

1.5. Conclusion

This chapter has highlighted the fact that the introduction of digital technology into our economies, far from being a simple technical change impacting the efficiency of production processes, implies systemic changes across sectors: modifications to the system of actors, public policies, knowledge and so on. Consequently, digital technology raises various questions about the development of technologies, their uses and their consequences. In agriculture, digital technology offers hope for its ability to address the multiple challenges of contemporary agriculture. It is therefore promoted in the discourse of various organizations, both public and private, national and international, agricultural and non-agricultural. Beyond being promoted in discourse, digital technology in agriculture mobilizes resources, whether for research, for funding startups in the field, for financing public-private projects or for encouraging investment in farms. However, the development of digital technology in agriculture is subject to controversy. These issues raise questions about the economic, social and political effects of digital technology, but also challenge its ability to address the environmental challenges facing agriculture. Reflecting on these controversies leads to two observations regarding the state of knowledge in this field. First, the information available on the uses of digital technology in agriculture, and even more specifically in French agriculture, is very limited. While some studies on the adoption of digital technologies exist, a systemic perspective is lacking, one that integrates digital uses into the socioeconomic environment of farmers and their production systems. Studies often view the development of digital technology in agriculture as a linear and uniform process, paying little attention to the heterogeneities and diversity of trajectories. Secondly, little knowledge is available on the interactions between the digitalization and ecologization of agriculture. The impacts of digital uses in real-world contexts are poorly documented, and there is very little literature that examines the structural impacts of digital uses on farms. This paradox, between a significant promotion of digital technology, on the one hand, and controversies that the current state of knowledge does not allow us to answer to, on the other hand, leads us to our empirical problem.

How does the development of digital technology in the agricultural sector impact and integrate into different ecologization trajectories? Does it favor some more than others?

The general overview presented in this chapter raises two major implications for addressing this problem. To compensate for the lack of knowledge regarding the development of digital technologies in agriculture, their uses and their implications, it appears necessary to conduct significant empirical research on this issue, integrating the microeconomic scale of these uses into a broader study of transformations within the agricultural sector. To integrate the various dimensions of digital technology development in agriculture and the controversies surrounding their effects on the sector‘s ecologization, a systemic and institutional perspective seems essential. To address these implications, we develop an analytical framework suited to a systemic, multilevel, heterogeneous and dynamic view of innovation by proposing a political economy of the agricultural innovation system.

2

Political Economy of Innovation and the Study of Agricultural Transformations

Chapter 1 demonstrated that digital development impacts several dimensions of agricultural activity, appears to lead to systemic transformations within the sector and is the subject of controversy and questions regarding its effects on different pathways to the ecologization of agriculture. Digitalization and ecologization can thus be viewed as sectoral innovation processes that unfold both within the institutions or organizations that frame this sector and at the level of farms and farmers' activities and practices. Indeed, these processes involve changes not only in farmers' practices but also at the sector level, in public policies, standards, knowledge development, advisory and learning mechanisms, and so on. To jointly study these different dimensions, an analysis focused solely on the adoption of digital technologies at the level of agricultural businesses seems insufficient. This chapter will present the theoretical framework on which this work is based, which combines these different dimensions by proposing a political economy of the agricultural innovation system (AIS), considered as heterogeneous, in which the emphasis is placed on the different scales of analysis of the sector.

The ecologization and digitalization of agriculture are processes studied by various approaches in agronomy and the social sciences. They can be analyzed as innovation processes since they correspond to the introduction of new elements into agricultural systems (new objectives, new technologies, new practices, etc.). Innovation economics thus emerges as a relevant initial general analytical framework, bringing together economic research that theoretically and empirically studies innovation processes, with different hypotheses or schools of thought. This framework partially overlaps with "Innovation Studies" (Martin 2012), which gathers works from several disciplines (economics, political science, management and geography) and shares systemic and dynamic views of innovation. Among the different conceptions of innovation briefly presented in this chapter, this work is based on an innovation systems approach, put forward by economists within innovation studies. This conceptual framework allows us to understand the processes of digitalization and ecologization of agriculture through different dimensions (knowledge, institutions, actors and technologies) and at different scales (national, sectoral and territorial). More specifically, we develop an institutionalist, evolutionary and political approach to innovation systems. Institutional economics places institutions at the heart of economic analysis by studying their nature, role, origin and the way they change (Chavance 2012). Institutions, which are understood as "autonomous collective entities contributing to the regulation of the socioeconomic system" (Walliser 1989, p. 339), are not considered as a contextual element, but rather as endogenous components of the economic system. They constitute a set of coordination mechanisms that extend beyond the market alone. Evolutionary economics focuses on the study of transformations in societies and institutions. Meanwhile, the literature on the ecologization of agriculture suggests considering the AIS as heterogeneous, comprised of several paradigms that co-evolve and establish different ways of envisioning the ecologization of agriculture. Political economy is then mobilized to integrate this heterogeneity, along with the resulting representations and power dynamics. This heterogeneity is found in the organizations and institutions of the agricultural sector. However, it is also embodied in the practices and production models at the farm level.

The aim of this work is therefore to analyze transformations at the individual and sectoral levels, which mutually influence each other and thus contribute to the construction of sectoral trajectories. These theoretical choices will lead to the central question that this book attempts to answer.

Section 2.1 explains the choice of a systemic approach to innovation to answer the research question based on a cross-analysis overview of theoretical frameworks of innovation and research on the digitalization of agriculture. Section 2.2 details how the heterogeneity of the agricultural sector and its ecological farming practices are integrated through the concept of paradigm. Section 2.3 revisits how to integrate multiple scales of analysis, adding the microeconomic scale through the analysis of uses and specifying the mechanisms of interaction between individual and sectoral levels.

2.1. A systemic approach to digital technology in the agricultural sector

2.1.1. *Digital development in agriculture: an innovation process*

The overview of digital technology in agriculture presented in Chapter 1 demonstrates that its development has a systemic dimension. Digital technology primarily refers to a material and technical dimension. It encompasses technologies with significant physical and functional heterogeneity. Digital technologies include diverse objects and processes and are often described as generic: they can fulfill a wide variety of functions. Digital technology also refers to uses and practices in design, manufacturing and use. It has a political and economic dimension, as it will impact production models and social and institutional dynamics. These technologies exist through a diversity of economic models, some based on the commercialization of physical artifacts, others on the commercialization of data or services and others through using free and open-source software. The development of digital technology is driven by a diverse range of actors, including digital companies, startups, as well as public and non-profit organizations. It thus involves inter-sectoral economic dynamics. Public policies and regulatory mechanisms impact digital development through economic policies, research orientation, and rules, but also various policies that can encourage or restrict the use of these technologies. This technological development is part of the historical dynamics of capitalism and its technological trajectories. Digital technology is developing in a range of sectors, including agriculture.

Thus, beyond a technical evolution, the development of digital technology is a trajectory:

– Multi-scalar: it plays out both in uses and in the broadest sectoral developments. It is not a niche dynamic, but a general and cross-cutting dynamic. The development of digital technology thus plays out at both micro- and meso-scales.

– Multifactorial and multiactor: it notably involves market developments, public policies, knowledge and strategies of actors more or less specialized in these areas.

– Intersectoral: with actors and institutions from the agricultural and digital sectors interacting.

– Dynamic: it involves changes in practices, actors and knowledge. It can modify the very structure of the agricultural sector.

Thus, the development of digital technology in the agricultural sector does not correspond to an external or natural technological evolution that transforms economies. It is an endogenous evolution within a sectoral economic system, driven by the decisions of economic actors and by institutions that guide, coordinate or determine it. The existing socioeconomic regime will impact the type of knowledge developed, the design of technologies, their dissemination and their functions. These developments can therefore be analyzed as social and institutional constructs. However, considering these technological developments solely as a social construct means losing "the ability to account for the differentiated fates that technological objects undergo" (Akrich 1989, p. 32). Technology is constructed and constantly interacts with social, technical, economic and cultural dimensions. The concept of innovation allows us to take the fluid and interactionist nature of technological change into account. The literature on the economics of innovation thus offers relevant concepts for analyzing the transformations of agricultural systems through digital technology.

> Innovation refers to the introduction of something new into a socioeconomic system, whether technological, product or process related, organizational, commercial, financial, social, etc. (OECD 1997). Innovation characterizes both the outcome – that is, the new element integrated into the system – and the process by which this integration occurs. It differs from invention, which Rogers (1962) defines as "the process by which a new idea is created or developed", whereas innovation is "the process of adopting an existing idea". Innovation can also take the form of new combinations of existing elements (Edquist 2013). By definition, innovation is an interactive process since it depends on the market, users, context, knowledge, resources, etc.

Box 2.1. *What is innovation?*

At the farm level, technological innovations can occur through the introduction of new tools such as digital technologies, as well as agronomic, environmental, organizational and commercial innovations. One type of innovation is often associated with other types of innovation: for example, the introduction of a new tool can change agronomic practices or the organization of the farm. Agricultural innovation is a process that combines technological, social, economic and institutional changes (Klerkx et al. 2012; Faure et al. 2018).

Different conceptions of innovation and different theoretical approaches to studying it exist and are presented in this section. Some approaches focus on the adoption of innovations by independent individuals. Other approaches study the design process of an innovation, from its emergence to its integration into socioeconomic systems. Some works consider innovation from a structuralist perspective, resulting from economic, social and political dynamics. Systemic approaches analyze innovation in its interactions with the economy and society: they jointly consider the institutions, organizations, networks, actors, knowledge and technologies that impact and are impacted by innovation.

2.1.1.1. *Advantages and limitations of methodological individualism for characterizing digital adoption*

An innovation can be studied through one of the stages of the innovation process: user adoption. This perspective relies on the notion of methodological individualism, that is, a conception of individuals acting according to their objectives, their optimization criteria, and their limited access to resources. The overall phenomenon of innovation diffusion is then understood as the aggregation of individual behaviors (Cusinato 2007). The institutional and political system is seen as an external factor influencing individual behaviors. Research then seeks to study the individual determinants of adoption in order to explain the social phenomenon of innovation diffusion within a given institutional context. The implicit assumption underlying this research is that technology adoption is linear and follows an "S-curve", as described by Everett Rogers in 1962 (Rogers 2010). Thus, the beginning of this curve would correspond to the rapid adoption of a new technology by a small portion of the population, the *pioneers*. Next, the slope of the curve would increase exponentially with the adoption of the technology by the early majority, then slow down with the late majority. Finally, the curve would flatten out with the late and slow adoption of the technology by the *laggards*.

The economic literature on digital technology in agriculture is primarily based on this type of approach: studies examine the adoption and factors influencing the adoption of technologies (Barnes et al. 2019; Lowenberg-DeBoer and Erickson 2019). This research aims to understand why some farmers adopt a digital technology while others do not.

The determinants of adoption vary depending on the technology studied (Konrad et al. 2019):

– Barnes et al. (2019) show that the adoption of precision agriculture technologies depends on the farmer's age, household income and farm size.

– Konrad et al. (2019) show that the adoption of precision fertilization technologies depends on farm size, propensity to innovate and the farmer's age, whether they are full-time or part-time on the farm, but also on the production system: presence or absence of livestock or organic farming, use of or absence of manure and the regulation in place on fertilization.

– Michels et al. (2020) show that smartphone adoption by farmers depends on the farmer's age, skills and innovative behavior, as well as the size and location of their farm.

– Torrez et al. (2016), for a set of digital farming technologies, highlight the role of the farmer's age and risk aversion, farm size, economic results and the length of time the farmer has been established.

Actors regularly refer to the Rogers curve and this individual vision of innovation, for example, by describing farmers as "early adopters", or on the contrary by pointing out a "delay in French agriculture" (Isaac and Pouyat 2015; Martin 2019).

Box 2.2. *Factors influencing the adoption of digital technology in agriculture*

While the factors vary from one technology to another and from one study to another, this research shows that technology adoption depends on:

– individual factors such as age, propensity to innovate, attitude towards risk, etc.;

– economic factors such as income and investment capacity;

– information available on innovation;

– the type of production system;

– the context and networks of farmers regarding the adoption of technologies.

Thus, membership in a cooperative or a farmers' group or guidance from an advisor can also influence the adoption of precision agriculture technologies (Barnes et al. 2019; Konrad et al. 2019). This research shows that existing environmental regulations and geographical characteristics also influence adoption.

These approaches are useful for highlighting a number of "conditions for technology adoption", understanding disparities in the diffusion of innovations and uncovering unanticipated barriers to adoption. However, these approaches leave blind spots regarding certain components of innovation phenomena. First, the innovation environment, social interactions and institutions can only be considered as an exogenous context. Innovation is both an individual and a collective process, where individuals and the system in which they operate interact and co-evolve (Hekkert et al. 2007). Individualistic approaches fail to capture network effects, at best limiting them to "egocentric networks" without the possibility of structural analysis. Furthermore, approaches based on technology adoption do not allow us to understand the uses and reconfigurations of technologies, which may differ from anticipated uses. To study the effects of introducing an innovation, such as the effects of digital technology on ecologization, it is crucial to go beyond simply adopting a technology. This is all the more important because technologies are not independent of one another, but combine with each other and with users' practices (Clapp and Ruder 2020). Furthermore, "adopters" are not passive users in the face of the arrival of external innovations, but are part of a network of actors and practitioners involved in the construction of innovations (Glover et al. 2019).

The phenomenon of innovation cannot therefore be understood solely as an aggregation of individual behaviors, but as a complex reconfiguration of social and technical components. Innovation will thus not be considered here from a methodological individualism perspective, but from an institutional economics perspective, even though approaches developed by adoption economists can be used as heuristics in a more global approach.

2.1.1.2. *Economic approaches to innovation as a process*

A technological innovation can be studied from another perspective that aims to understand its development process and the diversity of interactions that constitute it. These approaches thus study the different stages of innovation development and emphasize the processes, dynamics and temporality of innovation.

Before the 1950s, innovation was a concept largely absent from economic literature, with the exception of Schumpeter's seminal work. He placed innovation at the heart of economics, defining it as a modification of the production function that drives economic and social change (Schumpeter 1935). However, the innovation process itself was little studied, being primarily viewed as a result of entrepreneurial activity. It was only a few years later that innovation, as a process, became a significant subject of study, particularly with the aim of incorporating economic, political, managerial and organizational elements to promote innovation (Martin 2012).

Process-oriented approaches are used to study the design processes of digital agricultural technologies. Bronson (2019) shows that in the design of big data–based technologies in North America, designers rely on a simplistic view of farmers as rational agents seeking to maximize their income by producing staple crops on large farms. Maximizing yield and profit is central to the design of these technologies, which could explain their low adoption rate among farmers. Some studies focus on the actors involved in this process and show the growing presence of non-agricultural industries and phenomena of vertical integration of digital services, as well as the importance of public support (Eastwood et al. 2017; Birner et al. 2021). Oui's (2021a) work shows how the material and economic frictions encountered by companies during the design and commercialization process of an innovation lead them to deviate from their initial objectives. The work of Jakku and Thorburn (2010) and Ayre et al. (2019) demonstrates that the participatory design process for decision-support tools facilitates dialogue, cooperation and social learning among the stakeholders involved (farmers, advisors, researchers or industry professionals), which can lead to a better understanding of each other's challenges and changes in attitudes and practices.

These analyses help explain why and how innovations emerge and spread, or fail to do so. However, they offer little insight into the uses and effects of these innovations. While useful for understanding the emergence and success of certain digital agricultural technologies or companies, they are limited in their ability to study the transformations of the agricultural sector and its production models.

In economics, the innovation process was initially understood as a simple linear process: the invention arises from basic scientific research followed by applied research, is developed, produced and then brought to market and adopted by users (Martin 2012). From the firm's perspective, technological innovation is thus an exogenous factor of production (Solow 1956). The linear *push* model, in which innovation arises from science, has been adopted in numerous academic works, but especially in operational documents and guides. Other studies propose a different vision of innovation, also linear, but driven by demand, by the market: this is the *demand-pull* model (Schmookler 1966). The linear view of innovation has been criticized, both for its limited heuristic value and for the transformations it has fostered (Forest 2014). On the one hand, field studies show that the correlation between scientific research and innovation is not systematic, or that feedback or bifurcation phenomena occur in the development of an innovation, demonstrating that innovation does not necessarily unfold in a linear fashion (Griliches 1979; Scherer 1983). On the other hand, this theoretical vision has influenced public research and innovation policies since the 1950s, with controversial effects (Forest 2014). These innovation policies, for example, supported the Green Revolution, which refers to a set of research programs and the transfer of Western agricultural technologies (varieties, fertilizers and pesticides) to the rest of the world, criticized for its environmental, social and political effects (Glaeser 2010; Cornilleau and Joly 2014). To improve the understanding of innovation processes, Kline and Rosenberg (1986) proposed the interconnected chain model, in which design remains the central process, but innovation results from multiple interactions and feedback loops. Other critical and multi-disciplinary approaches have developed, proposing innovation concepts based, for example, on experience and local empirical know-how (Darré et al. 1989; von Hippel 2006).

Box 2.3. *Linear approaches to innovation*

2.1.1.3. *Innovation as a historical process*

Evolutionary economics has also addressed the question of the innovation development process, providing and theorizing a historical perspective (Nelson and Winter 1982). Research shows that the development of technological innovations is linked to existing technologies, knowledge and learning effects, as well as institutions. Innovation is therefore not an external process, but rather is embedded in technological trajectories. The seminal work of David (1985) and Arthur (1989) demonstrates the importance of studying the dynamics of the technological change process. They show that time can affect the value of a technological choice: this value can increase with the degree of adoption of that technological choice; this is what is known as "increasing returns to adoption". This phenomenon can lead to technological lock-ins, whereby a technology remains dominant even when a more efficient alternative exists. Evolutionary economics is therefore interested in the economic mechanisms that influence the success or failure of one innovation among others. Research shows that there are selection mechanisms that favor one innovation over another.

These issues of lock-ins, irreversibilities and selection are also addressed by other institutional approaches (Boyer et al. 1994). These approaches engage with evolutionary theories, which share the common ground of understanding economic change as a historical process rooted in institutions and of placing knowledge production mechanisms at the heart of these dynamics (Coriat and Dosi 2002; Labarthe et al. 2021b). Technological change must then be understood according to a principle of historicity (Boyer 1989). Existing institutional arrangements can lead to institutional lock-ins (Pierson 2000). The stability of an economic configuration can thus be linked to sociotechnical lock-ins: various externalities, specific infrastructures or institutions, as well as the acquisition and exploitation of skills, will reinforce the processes of lock-in and irreversibility (Kirat 1991). Beyond simply fixing technological trajectories, institutions can also be a source of exclusion for alternatives or of creating new techno-economic possibilities (Godard and Salles 1991). The direction of technological change will then depend on the existing power dynamics (Dockès 1990). The economic system, the technological system and institutions thus form a selective environment. The innovation process is therefore a process of technological and institutional co-evolution (Perez 1983).

This research highlights the importance of history for understanding economies: past events and choices influence present choices; this is known as "path dependency". In agriculture, for example, Cowan and Gunby (1996) show that the development of pesticides was driven by the knowledge about chemicals developed during World War II, as well as by the yield-maximization objectives of the time. The authors explain that, although criticized for its negative externalities, chemical pesticide-based control tends to remain dominant over integrated pest management (IPM) techniques (which aim to minimize economic losses by combining preventative, biological and chemical methods) due to coordination problems, reliance on other technologies and investments (such as treatment equipment), a lack of scientific and empirical knowledge about these techniques, and also due to public policies, perceptions and existing economic systems. This "path dependency" is true at the sector level, but also at the firm level, whose evolution will depend on the knowledge base acquired (Dosi 1982).

2.1.1.4. *Structuralist approaches to technological change*

Structuralist approaches to innovation study the influence of social, economic, political, institutional and cultural models on innovation within societies. These structuralist approaches to transformations in the agricultural sector developed in the 1980s and 1990s, within a context of increasing globalization, deregulation by states, international economic policies and international agreements (Buttel 1994).

Regulation theory, for example, proposes to study modes of production and accumulation according to the institutional forms of economies, that is, the codification of fundamental social relations (Boyer 1989). In capitalist economies, modes of development evolve according to a profit-driven logic. "Technological styles", defined as the most efficient and rational way of taking advantage of the cost structure, succeed one another and impact modes of development (Perez 1983). Innovation is viewed as a social production that depends on institutions, and in particular on forms of social relations, the political and cultural context, and modes of behavior (Dockès 1990).

These perspectives have notably highlighted the links between agricultural transformations and broader political and economic developments such as the Fordist regime. The book *The Great Transformation of Agriculture* shows that agricultural innovation in the second half of the 20th century, rooted in the Fordist economic dynamic, was then focused on increasing productivity, specialization and labor intensification (Allaire and Boyer 1995). Research in regulation theory demonstrates the role of labor relations, the state, the institutionalization of co-management between the state and agricultural unions, of the market, and of competition in the transformations of the agricultural sector (Labarthe and Touzard 2016). This research has also studied crises and associated transformations in agriculture, such as the shift toward a quality-based economy (Allaire 2002). "Food regime" approaches also offer a macro-historical analytical framework for studying the agri-food sector (Friedman and McMichael 1989). This research identifies three periods of stability in the agri-food system, associated with political, commercial and technological conditions: British hegemony from 1870 to 1914, American hegemony from 1914 to 1970 and then neoliberal hegemony. This analytical framework is considered less relevant today due to a diversity of poles rather than a single hegemony, which centers the analysis on the West, and is criticized for its homogenizing approach, which fails to show the diversity of existing trajectories (Wilkinson and Goodman 2018).

Box 2.4. *Structuralist approaches in agriculture*

NOTE.– The integration of digital technologies in agricultural operations is an innovation phenomenon that depends on:

– adoption factors: individual characteristics, economic factors, production system, networks and social context, available information;

– the technology design process;

– the trajectory, on past choices, both at the scale of the operation and at the scale of the sector (technologies already used, knowledge available, etc.);

– institutions: market organization, regulations, social relations, political and cultural context, etc.;

– the use and implementation of these technologies: adaptations, learning, appropriation, etc., which contribute to building innovation and influence the effects of this innovation.

These structuralist approaches, while interesting from a historical perspective, are more difficult to apply for analyzing ongoing change. Regarding the digitalization of agriculture, however, we can point to the pioneering work of Steven Wolf (Wolf and Buttel 1996; Wolf and Wood 1997), which shows that the development of precision agriculture in the United States is part of a historical continuity of several trends: (1) the decline of the state's role and the growth of industry's role in the production and dissemination of agricultural information, (2) the concentration and reorganization of production systems toward coordination and control by upstream and downstream industries, and (3) the increasing importance of environmental justification. This work demonstrates how the privatization and commodification of information in precision agriculture leads to a concentration of assets and a reorganization of production according to the demands of industrial capital. More recently, the work of Prause et al. (2020) also examines the effects of digitalization on the existing food regime. This work shows that digital technology appears to reinforce the characteristics of the existing system, namely, financialization, the use of the ecological narrative as a justification, the control exerted by large upstream and downstream firms, and a growing divergence between small and large farms. This development could lead to more structural changes for existing actors (notably in distribution, with the rise of Amazon and Alibaba) and for labor through the automation of certain tasks. This research underscores the importance of placing the analysis of digital technology in agriculture within a historical and macroeconomic perspective. However, the work of Wolf and Wood (1997) concludes with the need to develop complementarities between global and micro-level approaches, particularly by studying transformations on farms and across multiple geographical areas. Micro-level approaches appear essential, particularly for studying the impacts of technologies, the role of local knowledge, the implementation of practices and technologies on farms, and their complexity (Buttel 1994). Studying innovation therefore implies taking into account both macro- and micro-social mechanisms.

2.1.2. *Transformation of the AIS*

The trajectory of digitalization develops within the economic and political context of the agricultural sector, but in turn influences the agricultural sector. It depends on the social and institutional context, but also in turn impacts society and institutions. It is not simply a linear process originating in science and flowing to farmers. A multitude of factors contribute to the construction of this trajectory: available knowledge, institutions, various technologies, farmers' practices, etc. Therefore, analyzing this innovation trajectory requires a systemic perspective.

2.1.2.1. *The innovation system as a concept*

The concept of an innovation system offers an intermediate perspective that takes into account the diversity of actors who may play a role in innovation – both from research and from the field – their interactions, their knowledge and the context in which they operate. The innovation system concept represents a systemic, multidisciplinary and historical approach to the innovation process (Edquist 2013). It is used to jointly analyze the institutions, organizations, networks and actors that interact and impact innovation in a given space (Touzard et al. 2014). These impacts on innovation occur through a multitude of actions: development, dissemination, use, design, promotion, training, etc. (Granstrand 2000). Thus, this system will guide the type of innovation; it can foster some while hindering others. This concept of an innovation system is developed in the interdisciplinary field of Innovation Studies and at the intersection of innovation economics and institutional economics.

This type of approach appears relevant for studying the digitalization of the agricultural sector. Indeed, studies on the subject have shown the importance of actors, research, knowledge and skills, institutions and public policies, governance mechanisms of innovation and power, markets and economic models, or even networks and learning (Busse et al. 2014; Fielke et al. 2019; Rijswijk et al. 2019; Klerkx and Rose 2020).

In this work, we will therefore have a systemic perspective on innovation agriculture as a process of interaction between multiple factors and multiple scales, within a socioeconomic system that structures power relations.

2.1.2.2. *Various scales of the innovation system concept*

2.1.2.2.1. Territorial approaches to innovation

The innovation systems approach was initially used to analyze national differences in innovation dynamics, impacting economic growth and linked to differences in organizations and institutions (Freeman 1987; Lundvall et al. 2002). The diversity of research structures, research and development budgets, universities, economic regulations, routines, etc., induces diverse innovation dynamics. The authors thus developed the concept of a national innovation system to characterize all the national mechanisms that play a role in innovation and to compare these mechanisms across several countries. Beyond institutional and organizational differences between countries, these approaches emphasize the spatial dimension of innovation at subnational scales. The concept of an innovation system has therefore been applied to various territorial scales, such as industrial districts, localized production systems or innovation clusters (Torre and Tanguy 2014). These territorial approaches highlight the importance of local conditions for innovation, particularly the role of tacit knowledge exchange, trust and network externalities (Koschatzky et al. 2009; Torre and Tanguy 2014). These ideas have led to innovation policies such as the establishment of innovation clusters in France. These approaches could be used to compare the development of digital agriculture in different countries or territories. This would allow us to understand the role of institutions specific to each country or territory in the development of digital technology.

2.1.2.2.2. Corporate Innovation System

A new concept has emerged with the development of firms that have implemented innovation policies: the Corporate Innovation System (CIS) (Granstrand 2000). Indeed, some firms build networks that they dominate, allowing them to direct innovation according to their own needs (Lundvall and Rikap 2022). The innovation systems of very large firms interact and co-evolve with national innovation systems: both through processes of mutual reinforcement, as well as power struggle and competition. The introductory paper on agricultural big data by Bronson and Knezevic (2016) presents the fact that firms such as John Deere and Monsanto play a central role in the digitalization of agriculture and influence the direction of this process. This invites further research into the role and positioning of these firms within this process. However, this type of approach does not examine the uses and effects of innovations.

2.1.2.2.3. Sectoral innovation system

Research on innovation systems has also focused on sectors, which can be defined as "a set of activities unified by product groups, which respond to an existing or latent demand, and which share common knowledge bases" (Malerba 2004). Sectors differ significantly in terms of economic structures, the nature of innovations, knowledge sources and types of production, which implies very different innovation regimes (Pavitt 1984). Innovation is generally influenced by the characteristics of a sector and will in turn potentially transform or even create a sector of activity (Touzard 2014). The concept of a sectoral system of innovation (SSI) is used to analyze sectoral innovation within sectors from a multidimensional, integrated and dynamic perspective (Malerba 2004). This system will impact the innovation of firms and other actors, which in turn impact the organization of the SSI: indeed, organizations are considered active agents that participate in building the technological environment and the market. In particular, the emergence of new entrants or alliances with firms from other sectors can modify the sectoral innovation system (Temri 2014). Thus, the boundaries of the SSI are central to analysis using this concept, since they will evolve over time (Malerba 2004). Some approaches analyze the effects of sector institutions and networks on innovation processes, economic dynamics and firm strategies. Other approaches start from innovation processes and actors involved to identify the institutions and networks that impact them (Touzard 2014). The SSI concept highlights the role of institutions, firms, but also other organizations, market interactions as well as non-market interactions, and considers the sector as dynamic and not as constant data.

These different approaches to innovation can be combined since, for example, sectoral factors will interact with territorial factors, or technological innovation systems will interact with sectoral systems of innovation (Stephan et al. 2017). These different approaches are based on an interactionist view of innovation, where a multiplicity of actors, knowledge, institutions, networks and the environment interact.

2.1.2.3. *The agricultural innovation system*

The innovation system concept has been specifically adapted for agriculture (Klerkx et al. 2012). The agricultural sector has particular characteristics that justify a sectoral approach to innovation. First, the agricultural sector has been the subject of specific public policies and innovation support mechanisms, initially linked to the challenge of feeding the population after the Second World War. In Europe, national and European policies thus played a crucial role in the evolution of the agricultural sector in the post–World War II era. The work of Allaire and Boyer (1995), for example, shows that the technological change associated with Fordism spread to agriculture without affecting production relations in the same way it had in industry or the rest of the economy due to specific institutional arrangements. Among the specific characteristics of the agricultural sector, we can also mention the existence of dedicated research and training organizations. Furthermore, the sector's organization is unique, with highly concentrated upstream and downstream actors, while farms themselves form a collection of small, fragmented businesses. Collective action is also significant in this sector, which has its own institutions and a specific knowledge base. Added to this is the sector's unique relationship with nature, as well as its production of common goods, which raises new and major challenges such as biodiversity conservation and combating climate change (Touzard et al. 2014). Considering these specific characteristics has led several authors to propose the concepts of AISs or agricultural knowledge and innovation systems (AKIS) (Hall 2006; Klerkx et al. 2012).

Fielke et al. (2019) developed the concept of DAIS (digitalization of agricultural innovation systems) to characterize the "increasing influence of digital technologies on agricultural production, currently managed by networks of actors and technologies within traditional agricultural innovation systems". The idea behind this concept is to study digital development not as a set of independent and exogenous technological innovations, but as a set of interrelated innovations interacting with the context in which they are developed. Digital development will depend on actors, values, technologies, public policies and knowledge of the sector. And this development will, in turn, impact these elements. In this first article, they show that public policies are influenced by, and in turn influence, digital development. The evolution of advisory systems is detailed in another article (Fielke et al. 2020).

NOTE– This book will examine digital development through the lens of the agricultural innovation system concept. Innovation in the agricultural sector depends on actors and their interactions, as well as on knowledge, technologies, institutions and public policies. Digitalization transforms the innovation system, for example, through the arrival of new actors from other sectors, or the evolution of practices and uses.

2.1.3. *Choice of a structural analysis of innovation systems*

A sectoral system of innovation can be understood through a variety of approaches: functional approaches, process-oriented approaches and infrastructure-oriented approaches (Klerkx et al. 2012). These approaches are summarized below.

2.1.3.1. *Functional approaches to innovation systems*

Some authors suggest focusing on the activities and functions carried out within the innovation system. Indeed, innovation depends on a number of functions that may or may not be fulfilled within the context of the innovation system. Hekkert et al. (2007) define several functions that interact during technological innovation: knowledge development and dissemination, entrepreneurial activities, research directions, market creation, resource mobilization and creating legitimacy. These functions interact with each other, either synergistically or in opposition. This approach allows us to identify the constraints and shortcomings of the innovation system in fulfilling these functions. Based on this approach, Eastwood et al. (2017) highlighted deficiencies in evaluation and reflexivity in the implementation of precision livestock technologies in Australia. This approach is primarily suited to purely technological innovation, focused on firms, and takes little account of the environment (Markard and Truffer 2008).

2.1.3.2. *Process-based approaches to innovation systems*

Other approaches study innovation in terms of processes associated with an innovation system, that is, by focusing on the different stages of an innovation's trajectory, from its emergence to its eventual widespread adoption in society. The transition studies framework arises from critical work on innovation systems, but does not claim to be part of innovation studies (Geels 2002). From a different disciplinary perspective, this approach has the advantage of integrating temporality and the interaction between multiple scales. However, it shows limitations in analyzing the roles and strategies of actors, their interactions and the distribution of resources among actors. Furthermore, while it allows us to understand the possibilities for the emergence of certain innovations in niches, this approach does not explain the dynamics beyond them (Markard and Truffer 2008). This approach therefore does not appear entirely appropriate for analyzing the dynamics of digitalization and ecologization, which today involve more global transformations and are not limited to niches.

The multi-level perspective, arising from a convergence of actor-network theory, sociology of science and technology, evolutionary economics, and Innovation Studies, has given rise to a multidisciplinary academic community: *Transition Studies* (Geels 2004). It proposes an analysis of innovation trajectories based on three levels: the niche, the regime and the landscape. These approaches focus particularly on the construction of innovation niches. These niches are more stable spaces for innovation, partially protected from market selection, for example, through subsidies or a niche market, in which innovations can develop in co-evolution with users. These protected spaces allow for the creation of so-called radical innovations, which may or may not subsequently develop and transform the sociotechnical regime as a whole. This innovation perspective has been very successful, both in the scientific literature and in the implementation of public policies, particularly in the agricultural sector (Bui et al. 2016).

Box 2.5. *Transition studies*

The evolutionary economic approaches mentioned earlier can also be linked to systemic and process-oriented analyses of innovation. Indeed, all these analyses integrate the institutions, knowledge and actors that interact to shape innovation trajectories (Dosi 1982). Evolutionary economics shows that this trajectory is the result of power dynamics between actors and depends on the system's infrastructure, institutions and technologies, as well as knowledge (Nelson and Winter 1982). More recently, research has introduced the concept of directionality, which refers to guiding the innovation trajectory in a defined direction (Hekkert et al. 2020; Klerkx and Begemann 2020). This concept is notably used to highlight the role of innovation on society and the environment. It implies the establishment of innovation governance and coordination between technological and institutional innovation. These approaches are therefore valuable for understanding the innovation trajectories of ecologization and digitalization. They allow us not only to analyze past and ongoing trajectories but also to consider how to design and decide on future trajectories (Klerkx and Rose 2020). The evolutionary approach, and more broadly the process-based approach to innovation systems, will thus be used to discuss the results presented in this book and to situate them within a temporal perspective.

2.1.3.3. *Structural approaches to innovation systems*

Structural approaches aim to study the structural components that interact within innovation systems. Malerba (2004) proposes several categories of key factors that interact in the construction and evolution of innovation: knowledge, technologies, actors, networks, institutions and demand. There are constraints and interdependencies linked to existing technologies, hence impacting innovation. All actors, whether individuals, companies, or other forms of organization or groups of organizations, have their own skills, beliefs, language, objectives and behaviors, which will influence innovation. Between these actors, communication processes, exchange, cooperation or competition are established, which will depend on institutions. Institutions can be defined as the set of societal constraints that structure behavior, whether formal (laws, legal texts, etc.) or informal (normative rules – values, rights, responsibilities – and cognitive rules – frameworks of thought, language, concepts, etc.) (North 1991). Institutions frame the way actors act and interact, and therefore have an impact on innovation. Demand, formed by a heterogeneous group of individuals and organizations, can influence the behavior of firms and also impact innovation.

The framework of sectoral innovation systems was developed specifically to characterize sectoral changes, that is, the transformations and evolutions of variables within a sectoral system (Malerba 2002, p. 258). Change is an integral part of sectoral systems. And this change does not only mean a quantitative increase in variables, but can also involve the modification of the variables themselves. The structural analysis of these variables allows for the analysis of fundamental structural changes, such as the arrival of new actors, the emergence of new skills, new organizational forms and new strategies. These structural changes will affect the functions and interactions within an innovation system. Moreover, this analytical framework allows for the analysis of transformations within a sector when these transformations involve dynamics that go beyond the sector's traditional elements, whose scope extends across several sectors and redefines the boundaries of the sectoral system of innovation (Malerba 2002, p. 259).

This perspective has been used to study the development of precision agriculture and digital technologies for livestock management in Germany (Busse et al. 2014).

This work highlights sector dynamics, with the renewal of the role of advisors, the importance of farmers in this process, particularly in formulating needs and feedback, the importance of public institutions and policies, particularly in financing these innovations, the crucial role of interactions between different types of actors and the insufficient consideration of existing technologies (for compatibility and technical constraints).

The transformations in the agricultural sector brought about by digitalization and ecologization are modifying the structural components of the innovation system (actors, knowledge, technologies and institutions) and are even redefining its boundaries. Structural analysis will therefore be prioritized, while also incorporating elements of process analysis. This will lead us to analyze the interactions between the structural components of the innovation system and the agricultural transformations under study.

2.2. Integrating sectoral heterogeneity to study ecological agriculture

2.2.1. *Integration of ecological issues in agriculture*

2.2.1.1. *Diversity of ecological agriculture paradigms*

Observations of the harmful environmental effects caused by certain agricultural practices have led to an increased focus on environmental issues within agricultural research. In response to these findings, researchers, as well as agricultural organizations and social movements, have proposed an analysis of the potential synergies between agriculture and the environment, from both agronomic and social perspectives (Wezel et al. 2009). Some authors identify two main proposals for an ecological transformation of agriculture: one based on optimizing conventional agriculture and the other on alternative, or agroecological, agriculture (Duru et al. 2015). The former would propose a reduction in the environmental impacts of agriculture through the optimization of its practices and the improvement of efficiency and productivity. The second approach proposes an agricultural system based on the use and optimization of ecosystem services in order to benefit from the positive effects of the environment on production while limiting negative environmental impacts, or even generating positive effects. In reality, a wide variety of proposals have emerged to address environmental challenges, more or less explicitly linked to social, economic and political issues (Ollivier and Bellon 2013; Gasselin et al. 2021). For example, there is the development of soil-conserving practices in large-scale crop production, grazing and food self-sufficiency in dairy farming, as well as the valorization of local terroirs through appellations of origin in viticulture (which can be seen as a form of agroecology, due to its preservation of natural heritage and a production logic that is not solely focused on quantity). Within organic farming itself, a diversity of approaches is evident. On the one hand, various labels add further criteria for certification (Demeter, Nature et Progrès, etc.) (Brechet and Schieb-Bienfait 2006). On the other hand, faced with the success of organic agriculture, more and more farmers have adopted this production model, following different objectives and rationales, encompassing a diversity of practices. This is a source of tension within the sector and society, and echoes the scientific debate around the conventionalization of organic agriculture (Darnhofer et al. 2010; Stassart et al. 2012).

These proposals are implemented in diverse agricultural forms or models, a concept we will return to in section 2.3.1.1. Debates and controversies exist regarding the capacity of these proposals to best address these environmental challenges, bringing these agricultural models into conflict or coexistence. Is it through improving the efficiency of conventional production methods? Through the widespread adoption of organic farming? Or conservation agriculture? Through an agroecological transition of production systems? Through the coexistence of multiple production models? Through the introduction of digital technologies into these different production models (Connor 2008; Aune 2012; Giles and Stead 2021)?

To integrate these different proposals and the associated agricultural transformation processes, we use the concept of ecologization. Ecologization can be defined as the increasing integration of environmental issues into public policies, knowledge, organizations and professional practices (Deverre and de Marie 2008; Lamine 2011). Choosing this concept allows us to bring together the various approaches aimed at integrating environmental issues, regardless of the objectives set, the means used, the effects measured or the time frame considered. The concept of ecologization thus incorporates a plurality of perspectives. It implies a change in production activities to integrate ecological constraints or opportunities.

This concept of ecologization therefore has multiple facets and dimensions. How then can we study the processes of ecologization agriculture? Different disciplinary approaches, at various scales, offer analytical frameworks that are presented in section 2.3.

2.2.1.2. *Conceptual frameworks for ecological agriculture*

In agronomy, research on ecologization seeks to characterize practices according to their environmental impact and the transformations they induce in agricultural production systems. Hill and MacRae (1996) proposed a model, called ESR, describing three possible stages of ecologization practices. The first stage, efficiency, involves improving the efficiency of processes in order to reduce and optimize the use of inputs. The second stage, substitution, goes further, as it involves replacing conventional inputs with more ecological ones, such as organic inputs. Finally, redesign signifies a more in-depth reorganization of the production system according to ecological principles. These strategies of efficiency, substitution and redesign overlap and combine in the ecologization trajectories of farms (Lamine 2011). Approaches based primarily on efficiency and substitution would correspond to what is called "weak" ecologization and those based on redesign are classified as "strong" ecologization (Horlings and Marsden 2011[1]). Strong ecologization can be linked to agroecology, a concept that proposes a redesign of agricultural production systems based on ecological principles, but also a more comprehensive redesign of agri-food systems (see Box 2.6).

Other characterizations of agricultural systems integrating agronomic and socioeconomic perspectives have developed in order to move beyond a simplistic binary division and to go beyond purely agronomic criteria. Therond et al. (2017) define three biotechnological models (based on chemical inputs, based on biological inputs, based on biodiversity), which are combined with four types of socioeconomic context. They thus identify six models that differ notably in their systems of social values, or justification registers from a conventionalist perspective (Plumecocq et al. 2018).

The organic farming versus conventional agriculture dichotomy remains institutionally relevant. On the one hand, organic agriculture is easily identifiable through its label and associated constraints. On the other hand, it has its own "knowledge base", specific actors, networks, public policies, organizations, modes of interaction, learning processes and specific modes of regulation (Morgan and Murdoch 2000; Ingram 2018). Moreover, organic agriculture embodies a project that resonates with current societal expectations and political trends toward reducing or even eliminating synthetic pesticides.

The social sciences emphasize that ecologization takes shape and is oriented through conventions that will in turn structure practices (Mormont 2013). These conventions may or may not involve rules, incentives, standards, individual choices, commercial developments, collective commitments and technological changes. The production of these conventions involves diverse actors: research and development organizations, advisory and training structures, public organizations and agricultural professional bodies, all of which will impact the technologies and knowledge

1 Behind these different conceptions of the ecologization of agriculture, philosophy associates different perceptions of "nature". On the one hand, nature is seen as requiring artificial intervention in order to create a suitable agroecosystem. On the other hand, nature is seen as requiring management in order to guide and work with natural processes (Larrère 2002). In philosophy, the concept of ecologization can be used in a more restrictive way than the vision proposed here in a view opposed to modernization (Latour 1995).

created and disseminated for ecologization (Stassart and Jamar 2009). Companies within value chains and sectors will also structure these different pathways to ecologization (Magrini et al. 2016). This raises the question of the decision-making processes that govern these conventions, and therefore the power relations that will influence these decision-making processes. Thus, ecologization does not only play out in individual practices, but also at other scales since it involves decisions and processes related to institutions, networks, technologies or even the development of knowledge[2].

Agroecology is defined as the application of ecological concepts and principles to the design and management of sustainable agri-food systems (Gliessman 1998; Francis et al. 2003). These systems' agronomic principles must enable biomass recycling, guarantee favorable soil conditions, minimize resource losses, promote genetic diversification, allow for biological interactions and synergies, and enhance agrobiodiversity (Stassart et al. 2012). Research on agroecology has linked these agronomic characteristics to systemic approaches that integrate the agri-food system as a whole, within an interdisciplinary perspective. First, this agronomic evolution requires new forms of knowledge and learning. Stassart et al. (2012) add methodological principles to this, such as the multi-criteria management of agro-ecosystems from a transition perspective, the valorization of spatiotemporal variability, the exploration of new situations far removed from already known local optima and the construction of participatory research mechanisms. Second, in order to integrate the various components of the agri-food system that interact with agricultural production systems, this concept incorporates socioeconomic, political and scientific dimensions. Agroecology has thus become a unifying concept between scientific disciplines, social movements and agronomic practices (Wezel et al. 2009). Agroecology therefore integrates principles of bringing producers and consumers closer together, as well as the autonomy of local communities, the creation of knowledge and of collective adaptive capacities, the creation of an environment conducive to public goods and the valorization of diverse knowledge (Stassart et al. 2012). The concept of agroecology has gained popularity and now encompasses a wide range of realities. Indeed, agroecology is not always defined in the same way, and this depends on the type of actor or the geographical area (Van Hulst et al. 2020). Thus, agroecology is not a single form of agriculture, but a diverse set of practices, values and objectives.

Box 2.6. *Agroecology*

2 The concept of ecologization is sometimes linked to the concept of ecological modernization, theorized in sociology by Spaargaren and Mol (1992). This concept can be considered a path to ecologization, which compromises between a neoliberal vision of the economy and environmental critiques (Mormont 2013). Although it is a dominant framework in the implementation of environmental policies, this concept corresponds to a specific vision of ecologization and does not encompass the plurality of proposed trajectories.

The ecologization of agriculture can thus relate to a gradual optimization of agricultural practices, a substitution of inputs, an agroecological transition of territories, as well as a more rapid and radical transformation of production models. Ecologization is a process that is largely endogenous to the agricultural sector. It involves diverse actors within the sector and supply chains, as well as both "indigenous" and new institutions and knowledge. It can be described as heterogeneous given the diversity of the trajectories proposed and implemented. This heterogeneity is structured by power relations between the actors who drive these different trajectories. Thus, the issue of ecologization, in its multiple scales and dimensions, raises questions about individual practices, but also about the institutions that stabilize power relations between actors.

2.2.2. *Political economy of AISs*

Innovation system approaches tend to minimize power asymmetries within innovation processes, in which the various stakeholders are considered non-hierarchical (Lundvall and Rikap 2022). However, within an innovation system, there exists a diversity of actors with sometimes divergent interests and capacities for action that are not necessarily equal. Power relations between actors influence system transformations and innovation, which in turn influences these power relations. Political economy integrates these power relations to study their role in innovation.

On the one hand, research in political economy has demonstrated the heterogeneity of existing institutional structures. Thus, there is not one single homogeneous capitalism, but rather a variety of capitalisms (Hall and Soskice 2002). Consequently, there is also a variety of associated innovation models, theorized by the concept of a social system of innovation and production (SSIP), which aims to highlight the interrelationships between innovation dynamics and production dynamics. This research shows that there are different categories of SSIP, depending on the type of institutions in place, their complementarities and their hierarchy (Amable 2000). Institutional configurations influence the direction of innovation via incentives and constraints that promote diverse objectives (social justice, liberalization, etc.) and change the role of actors in innovation processes (businesses, public institutions, etc.).

On the other hand, political economy emphasizes the heterogeneity of agents within socioeconomic systems. This heterogeneity leads to a divergence of interests, which can be a source of conflict. Institutions – that is, the rules that structure social interactions – result from strategic interactions between actors within a specific power structure (Amable 2005). Institutions are thus considered here as the stabilized expression of power relations or as the result of compromises between heterogeneous interests, and these depend on the actors' ability to influence institutional choices. In turn, institutions will influence the heterogeneity of agents by impacting the distribution of gains or access to resources. Work in regulation theory has thus shown that institutional arrangements regulate access to resources between different social groups, and will therefore influence sectoral transformations (Touzard and Labarthe 2017). In particular, these rules will affect the conditions for the creation and diffusion of knowledge, which are central steps in innovation processes. Thus, institutional arrangements will tend to favor innovations that correspond to dominant production models. For example, work in the agricultural sector has shown how the agricultural research system promotes the development of genetic engineering more than agroecological innovation (Vanloqueren and Baret 2009).

This question of power is addressed in various social science works on the development of digital technology in agriculture. Many studies show, in particular, how the current development of digital technology reinforces existing power relations between farmers and upstream and downstream firms, and accentuates the concentration of power (Lajoie-O'Malley et al. 2020; Lioutas and Charatsari 2020). The growth of these technologies could even lead to a transfer of power from farmers to firms, particularly regarding information and knowledge (Carbonell 2016; Carolan 2017a). These phenomena of reinforcement and transfer are notably linked to the appropriation of data and benefits by these firms (Rotz et al. 2019a; Prause et al. 2020). This power manifests in various forms such as manipulation, attractiveness, dependence (leadership) and even coercion or force (Ryan 2019). More rarely, these power issues are raised to describe the oppositions that exist between several approaches to agriculture in the digitalization process, which could lead to phenomena of exclusion and to the monopolization of forms of production (Barrett and Rose 2020). Power relations do indeed play a role in controlling the direction of technological utility, potentially favoring certain visions of sustainability over others (Clapp and Ruder 2020). But these questions remain very theoretical and are poorly informed by data on the actions, uses and practices of agricultural actors.

The digitalization and ecologization of agriculture can be studied from an innovation systems perspective to account for the different scales and dimensions involved in these processes. However, research on innovation systems tends to neglect internal heterogeneities within the system (Malerba 2002). This issue of heterogeneity is all the more important in the context of the ecologization of agriculture, a phenomenon described as heterogeneous in the literature. Adopting a political economy approach to sectoral systems of innovation allows us to integrate this issue of heterogeneity and power relations within these systems.

2.2.3. *Integrating heterogeneity within the agricultural sector*

Analyses in terms of sectoral systems of innovation have already made it possible to identify sectoral variability in the field of innovation. However, the analysis of heterogeneities within sectors remains one of the areas where these theories need further development, particularly to understand these systems' dynamics (Malerba 2002, p. 262). This section aims to clarify (1) how we integrate this system heterogeneity in order to approach it from a political economy perspective, (2) how this heterogeneity influences the trajectory of innovation systems and (3) how these phenomena manifest in the agricultural sector.

2.2.3.1. *Paradigm concept in capituring the system heterogeneity*

To the existing diversity of social systems of innovation and production corresponds a diversity of socio-technical paradigms, that is, ways of thinking about production, of conceiving problems and of defining the field of possible solutions (Dockès 1990). A technological paradigm is "a model or a 'pattern' of solution of selected technological problems, based on selected principles derived from natural sciences and on selected material technologies" (Dosi 1982). The techno-economic paradigm is enriched and takes market structures, production conditions, and distribution into account (Djellal 1995). The sociotechnical paradigm considers technological innovation "as part of a broader social innovation". A paradigm is a way of viewing problems and considering solutions. A sociotechnical paradigm therefore includes a socio-organizational dimension, a technical dimension, and an economic dimension. We can add a more cultural and symbolic dimension: relationship to nature, knowledge, space, society, worldview, etc. (Beus and Dunlap 1990). These sociotechnical paradigms guide the innovation types according to existing modes of production, available resources, and according to economic and institutional structures.

These paradigms succeed one another over time and can be transferred from one sector to another, from one territory to another. Paradigms can also coexist: despite a dominant paradigm, marginal ways of thinking about production (and its relationship to nature and society) persist (Dockès 1990). Coexistence is understood here as the recognized and intentional side by side existence of several systems that may be in competition, complementarity, or juxtaposition.

This heterogeneity within innovation systems is rarely studied, although it can influence the practices of actors and dynamics at play (Malerba 2002, p. 262). Indeed, this heterogeneity affects the trajectories of innovation systems.

2.2.3.2. *Paradigms and trajectory of innovation systems*

A paradigm becomes stabilized within its own specific functioning, which can be called a regime. A technological regime is a "set of rules embedded in practices, production processes, product characteristics, skills and procedures [...], themselves embedded in institutions and infrastructures" (Rip and Kemp 1998, p. 338). The concept of a sociotechnical regime adds the various social groups that interact to create and maintain this regime (Geels 2004). The interdependencies and alignments between the different components of the regime (knowledge, infrastructures, rules, etc.) give it stability (Geels 2002). A regime will evolve along a trajectory that depends on its paradigm. Indeed, changes, whether technological or institutional, will depend on frameworks of thought, existing knowledge, learning systems, forms of organization, available resources, norms, and innovations already implemented (Kirat 1991) In order to analyze this cumulative nature of change, evolutionary economic theories introduce the concept of technological trajectory.

A technological trajectory is a "given initial technological potential which, due in particular to the cumulative, specific, and irreversible characteristics of innovation processes, would develop in a number of directions or trajectories, but within the precise limits defined by trade-offs between technological and economic factors and with significant exclusionary power" (Dosi 1982). Thus, a technological trajectory will be characterized by multiple incremental innovations within the regime and paradigm. Each regime can be associated with a cognitive system. A cognitive system is a broad set of knowledge, norms, beliefs, and guiding images that stabilize a production chain (Stassart and Jamar 2009). Institutions and the cognitive system stabilize the regime, potentially leading to lock-ins. These lock-ins are accentuated by the power relations that develop between actors (van Bers et al. 2019). Dominant actors and organizations will frame the direction of change to maintain their interests, via constraints and dependencies, technological innovations, or their influence on regulations and institutions (Cleaver 2002; Conti et al. 2021). These mechanisms will exclude alternatives or render them invisible. Trajectories thus depend on power relations.

Various paradigms and technological trajectories coexist within innovation systems. This coexistence implies conflicts, negotiations, and compromises. Indeed, these paradigms can compete or, conversely, act in a complementary manner. These interactions between paradigms lead to evolutions and can influence the directionality of innovation systems. On the one hand, these interactions define the relevant problems or needs that innovation must address (Kirat 1991). The legitimacy of actions and their social justification have been recognized as impacting economic dynamics (Amable 2005). This can lead to the evolution of existing paradigms. On the other hand, the interactions between paradigms influence how these identified problems are addressed. The interactions of these paradigms construct shared representations of desired futures and thus determine the types of solutions considered legitimate (Lajoie-O'Malley et al. 2020). This will impact the distribution of resources between actors and between paradigms and affect the deployment of innovations (Pigford et al. 2018; Bronson 2019). Alternative innovations can develop in niches and then potentially influence the regime later (Pigford et al. 2018).

For example, dominant paradigms tend to appropriate some of the positive aspects of alternative paradigms by accepting compatible components and eliminating the more radical implications (Beus and Dunlap 1990). Thus, addressing ecological challenges is no longer an argument solely put forward by alternative agriculture, but is widely adopted by conventional agriculture.

Conversely, innovations help to determine different trajectories, which can impact the paradigms themselves (Gasselin et al. 2021). Innovations are not neutral; they can have different impacts depending on their justification and form, which reflects a specific vision of nature, space, society, knowledge, territories, the market, the state, and technology.

Thus, the trajectory of an innovation system will depend on the dynamics of emergence, coexistence, and confrontation between paradigms. The direction and type of deployment of innovations are the result of negotiations and power struggle, and can in turn impact the mechanisms of coexistence. The deployment of innovations depends on their compatibility with institutions (property rights, diffusion models, production structures, etc.). Therefore, innovation is an interactive process of technological development and institutional change (Hayami and Ruttan 1971).

2.2.3.3. *Paradigm heterogeneity and models within the agricultural sector*

Various approaches to ecologization agriculture are proposed and implemented in agricultural practices, but also in marketing methods, knowledge development, and institutions. Several paradigms therefore coexist within the agricultural sector and innovation system (Gaitán-Cremaschi et al. 2019). Thus, there would be a conventional agrochemical paradigm, based on the simplification of systems, the use of inputs, specialization, intensification, and economies of scale to produce at low cost – a paradigm that became institutionalized and dominant after the Second World War (Labarthe and Touzard 2016; Magrini 2018). There would also be an agroecological paradigm, based on the use and amplification of ecosystem services to manage soil fertility, plant and animal health, and ultimately, agricultural production. Beyond this production-centric view, several paradigms could be defined by including marketing types or connections to territory (Plumecocq et al. 2018).

The ecologization of agriculture encompasses a wide variety of possible pathways for farms, based on the implementation of more environmentally friendly practices and adapted methodological and socioeconomic principles. However, these individual practices are promoted, supported, documented, and valued at other levels. Individual practices are embedded in knowledge systems, institutions, public policies, and value chains. They are integrated into sectoral dynamics. Indeed, the various conceptions of agriculture reflect a diversity of cognitive systems (Stassart and Jamar 2009). For example, the organic farming sector requires a different knowledge system (Morgan and Murdoch 2000). This concerns knowledge about production mechanisms, but also about marketing and processing. The specific characteristics of organic agriculture in particular necessitate the incorporation of scientific knowledge into local knowledge and community learning mechanisms. Thus, the development of knowledge in the agricultural sector influences the types of agricultural practices implemented. Scientific production also reflects the various paradigmatic poles of ecologically oriented agriculture (Ollivier and Bellon 2013). Vanloqueren and Baret (2009) demonstrate that scientific research in agriculture can influence agricultural trajectories by hindering agroecological innovation. Specifically, the authors highlight how the continuity of the agricultural technological trajectory is favored by research orientations, partnerships, the role of the media and lobbying groups, intellectual property, capitalist dynamics, implicit assumptions of the public sector, and the organization and mechanisms for evaluating research. Public policies influence research directions and, consequently, certain knowledge systems. But beyond that, public policies and institutions contribute to the construction of a belief system (Stassart and Jamar 2009), favor certain agricultural structures over others, influence the technological equipment of farms, and encourage or constrain agricultural practices and types of production. This heterogeneity of knowledge and institutions is also found at the level of advisory, collection, and supply organizations, as well as professional organizations. Collective and support dynamics play a crucial role in the evolution of agricultural models, particularly through the sharing of experience, the search for solutions, and via the construction of professional recognition (Lamine et al. 2009). At the market level, there is also a certain segmentation by distribution organizations or by labels, notably the organic farming label, but also designations of origin or other labels based on environmental, social, geographical, or nutritional criteria. Following this line of research, several articles emphasize that the development of digital technology tends to reinforce the dominant industrial agricultural system (Wolf and Buttel 1996; Bronson and Knezevic 2016).

Paradigms are found in institutions, organizations, and infrastructures. The concept of paradigm allows us to recognize the diversity of principles and institutions specific to each form of agriculture and ecologization. However, these different paradigms evolve within the same sectoral innovation system: they are subject to the same regulations and agricultural policies, develop within a partly similar scientific and technological environment, and operate within shared institutional frameworks. The concept of paradigm seems relevant to our research question because it articulates socio-organizational, technical, and economic dimensions, and allows us to integrate productive microsystems (here, farms) and society as a whole (here, limited to the agricultural sector) (Dockès 1990).

In reality, the boundaries between these paradigms are blurred: a continuum of practices exists, and differentiation processes occur at each end of the spectrum. Therefore, there is no clear line between conventional and agroecological agriculture (Van der Ploeg et al. 2019). Alternative, sustainable, or agroecological farming encompass a diversity of phenomena built around a diversity of conceptions of quality (Sonnino and Marsden 2006). Even within the formalized ideal type of alternative agriculture, organic farming, we observe differentiation processes, such as the conventionalization of organic agriculture (Darnhofer et al. 2010). Conventional agriculture is also undergoing various transformations. For example, Lucas (2021) describes a phenomenon of silent agroecology, which corresponds to the development of ecological practices within so-called "conventional agriculture", but which are not characterized as such by farmers who avoid ecological terminology, and which are poorly reflected in agricultural statistics. Crop diversification or conservation agriculture, which involves implementing farming practices to protect soil potential by minimizing tillage, permanent soil cover and crop diversification are also developing in both conventional and organic agriculture (Goulet and Vinck 2012; Morel et al. 2020). An increasing number of innovations are being proposed to implement substitution practices, situated between conventional agriculture and agroecology, such as the development of biocontrol (Aulagnier and Goulet 2017). Alternative marketing practices, such as local distribution channels and farm sales, are developing on farms with both conventional and more alternative production methods. The development of digital technology in agriculture is sometimes presented as a new paradigm, a new agricultural model. Sometimes, digital technology is presented as being able to adapt to the diversity of existing agricultural models. Finally, digital technology is sometimes presented as a component of the conventional agricultural model.

Agriculture is therefore heterogeneous: different paradigms coexist. These are found in practices, but also at the level of organizations and institutions. These paradigms do not evolve independently, but are in constant interaction within the agricultural sector and its innovation system, which can influence technological trajectories and therefore, these paradigms' modes of coexistence.

2.2.3.4. *Interactions between paradigms in the agricultural sector*

Institutions and organizations structure power relations between paradigms, thereby influencing the dynamics of the agricultural system and shaping its directionality (Pigford et al. 2018). The coexistence of paradigms can result in processes of co-evolution, convergence, but also differentiation and divergence.

Heterogeneity within the AIS creates competition for resource mobilization, social recognition and market access. The coexistence of paradigms can lead to confrontations over the type of knowledge to be developed (Vanloqueren and Baret 2009). For example, Morgan and Murdoch (2000) showed that power relations between conventional and organic agriculture have influenced knowledge development to the detriment of organic farming. Thus, the programing of public or private research plays a role in maintaining certain agricultural paradigms. Beyond scientific knowledge, there can be competition over the place of cognitive systems (knowledge, norms, beliefs and imaginaries) specific to each paradigm in public discourse and policy (Stassart and Jamar 2009). Oppositions also exist in terms of values, philosophical positions and conceptions of the relationship with nature (Plumecocq et al. 2018). These conflicts between conventional and alternative agriculture are also found in scientific circles, leading to different results, as they are affected by the beliefs and values of those producing them (Beus and Dunlap 1990). Research, knowledge development and various forms of support for innovation can influence technological innovation. However, technological innovation, including digital technologies, can also favor certain paradigms over others, depending on the objectives, values, knowledge or ways of doing things that are integrated into the technologies (Bronson 2019). The power dynamics between paradigms also influence institutional forms, which can favor certain paradigms more or less than others.

However, forms of synergy or hybridization also exist within this coexistence of paradigms. First, the alternative paradigms make it possible to respond to the criticisms of the conventional paradigm, which may ultimately favor its preservation rather than a more fundamental questioning of it. By integrating components of the alternative paradigm, the dominant model appropriates positive images associated with the alternative paradigm while ignoring its more radical implications, thus allowing it to maintain its dominant position (Beus and Dunlap 1990). But synergies can also benefit the different paradigms. Functional complementarities can exist, such as the sharing of resources, knowledge, labor, markets, or value (Gasselin et al. 2021). Sources of innovation can be reused in other paradigms, potentially leading to forms of hybridization (Plumecocq et al. 2018).

NOTE.– Our political economy approach will be based on a conception of the innovation system as heterogeneous, a site of coexistence and confrontation between several paradigms of agricultural ecologization. These paradigms are found at the level of individual practices, but also at the level of organizations and institutions. These paradigms interact, and changes in the innovation system can affect the mechanisms of interaction, favoring hybridization, synergy or competition and confrontation.

2.3. Integrating multiple scales to study AISs

2.3.1. *Integration of the microeconomic scale*

The framework for analyzing innovation systems, while integrating users and demand, does not account for the specific role of uses in the innovation process (Hekkert et al. 2007; Busse et al. 2014). Promising theoretical links can be established between structural economic analyses of changes at the industry/sector level and microeconomic approaches to innovation processes at the individual and organizational levels (Dosi et al. 1988, p. 1162). Focusing on individual practices and changes at the microeconomic level allows us to study the materialization of global changes. However, individual practices are not viewed here as separate from more global developments. They are embedded in a socioeconomic system and are in constant interaction with it: they are therefore subject to the constraints of this system while simultaneously modifying it. Innovation is therefore a multiscale process, in which systems and practices-based approaches are complementary. This section aims to clarify how the micro-scale will be integrated into the analysis using agricultural model concepts, practices and uses, and why integrating this scale is essential in the study of innovation processes.

2.3.1.1. *Agricultural model for distinguishing between farm types*

To characterize the diversity of production microsystems associated with the paradigms, we use the concept of the agricultural production model. In institutional economics, the concept of the production model is used to distinguish between different types of enterprises according to the type of product, resources, organization and labor (Boyer and Freyssenet 2000).

The concept of a production model characterizes production systems according to three components. The first is the company's product policy. Production models are thus distinguished according to the target markets, the product range offered, sales volume objectives, diversity, quality, innovation and the target profit margin. The second component corresponds to the production organization. This includes the methods and resources chosen, the degree of integration of activities, spatial distribution, the organization of design, procurement, manufacturing and marketing, the techniques employed, and management criteria. Finally, the third component characterizes labor and the employment relationship. Models are then distinguished according to recruitment methods, employment, classification and employee representation. Productive models can be defined as "corporate governance compromises that enable the sustainable implementation of a viable profit strategy within the growth patterns of countries where firms organize their activities using means that are coherent and acceptable to the stakeholders" (Boyer and Freyssenet 2000, p. 33). This concept thus allows us to distinguish between different business models while integrating them into their economic context.

Box 2.7. *The production model*

In agriculture, three uses of the notion of an "agricultural model" exist (Gasselin et al. 2021). The first refers to the concrete system, modeled by observers, a definition that aligns closely with that of a production model. The second uses the agricultural model as a project or desired future, which is idealized to make it a reference point for legitimizing or contesting changes. The third use of the concept defines it as a normative framework (agricultural policies, specifications, rules) that results from power dynamics. Models differ in terms of their techno-economic organization, values, knowledge and links to the market, the state and the territory (Plumecocq et al. 2018; Gasselin et al. 2021). In this study, we adopt the concept of model as a concrete system, which we refer to as an "agricultural production model", consistent with the institutional perspective.

Applied to agriculture, the concept of a model integrates biotechnical characteristics, that is, agricultural practices, in addition to their socioeconomic characterization (Rémy 1987; Therond et al. 2017). Indeed, these biotechnical characteristics are part of the productive organization of the agricultural production model. The notion of agricultural model can be related to "farming styles", as defined by Van der Ploeg (1994), who distinguishes farm models according to the types of resources used and the ways these resources are combined. Agricultural models can be conceived as (1) a microeconomic unit, with production processes, flows and constraints; (2) a "managed system" subject to decision-making processes related to objectives, knowledge and organization; (3) a component of a social system; or (4) a complex organization that coordinates a set of activities (Laurent et al. 2003). Agricultural models describe farm production, organizational structures, labor and practices, considering farms as embedded within a broader socioeconomic system. Technical change on farms is considered here as a process of negotiation between individual, social, economic, institutional and political factors (Laurent et al. 2003). Agricultural models are therefore where agricultural paradigms materialize; they are the concrete production systems that are implemented, evolve under the influence of institutional factors and, in turn, influence these factors. Agricultural models can therefore be studied through the practices and uses implemented.

2.3.1.2. *Technologies, uses and practices*

As mentioned in Chapter 1, a technology is not simply a physical object, but rather a body of knowledge, know-how, methods, procedures and experiences. To grasp the various dimensions of technology, it is therefore necessary to examine its uses.

The concept of use allows us to consider technologies in their actual application and in interaction with their context. This concept addresses technology beyond its purchase and prescribed use, considering its recurring integration into production practices and systems (DiMaggio and Hargittai 2001; Badillo and Pélissier 2015). Actual use is used to define "the capacity and possibility of successfully integrating ICTs into the achievement of objectives defined by the individual themselves or in collaboration" (Gurstein 2007, p. 9). This notion of effectiveness has been added to works on digital technology to highlight the issue of individuals' ability to use digital technologies beneficially. Few initiatives are implemented to address this issue, with public policies focusing instead on the problem of access and the "digital divide". However, the use of technology is associated with knowledge building, adjustments and interactions with the various practices of users (Higgins et al. 2017).

This concept of use allows us, on the one hand, to consider the interdependencies between different technologies, but also the economic, social and cultural context, which influences uses. On the other hand, this concept allows us to move beyond a vision of passive users and understand how they arrange, tinker with and adapt technologies to meet their needs. The concept of practice includes that of use, while also incorporating the related behaviors, attitudes and representations (Millerand 1998).

Research in agricultural digital economics mainly focuses on the adoption of technologies. However, a growing number of studies in the social sciences go beyond adoption and highlight the importance of focusing on uses. One example is the work of Ayerdi Gotor et al. (2020), which demonstrates various recurring combinations of precision agriculture technologies. These combinations are linked to specific objectives and have impacts on farm work (e.g. an increase in night work). The paper by Rose et al. (2018) shows that the uses of decision support tools (DSTs) does not consist of a literal application of the tool's recommendations, but rather that users take a step back and consider this information, combining it with their own knowledge. The work of Visser et al. (2021) questions the "precision" of the uses of big data–based technologies. There appears to be a difference between the promise of precision and the actual precision linked to the concrete implementation of these technologies. This precision is not simply a consequence of misuse by farmers, but rather results from the entire socio-technical framework surrounding these technologies.

2.3.1.3. *The importance of the micro-scale in the study of innovation processes*

Innovation is not merely a macroeconomic dynamic; it also depends on the micro-individual conditions under which innovation can or cannot materialize (Fleming et al. 2018). Without these uses and practices, there is no innovation. The conditions for the implementation of technological and institutional changes are an integral part of the innovation process. In order to move beyond an idealized, deterministic or simplified view, in which technological and institutional transformations would have a uniform effect corresponding to the anticipated outcome, it is necessary to study the concrete effects, the diversity of uses and the realities behind a macro-level effect (Buttel 1994). This allows us to study how innovations interact in practice, as is the case with, for example, the use of digital technology and ecologization.

Concrete practices and uses may differ from those envisioned. There may be unforeseen rearrangements, repurposing, lock-ins, etc. For example, the difference in vision of the farming profession between practitioners and developers of decision support tools leads to uses of these technologies that differ from those anticipated by their designers (Rose et al. 2018). Similarly, Wolf and Wood (1997) show that in practice, precision agriculture fulfills different functions from the economic and ecological claims it makes. In another area, the article by Hennessy et al. (2016) illustrates the difference between expected and observed effects, showing that farmers' use of the internet does not address the problem of social exclusion, since isolated individuals use the internet less. It is also in practice that phenomena of adaptation, resistance and the construction of alternatives will play out. This can be illustrated with the case of Farm Hack, an association of farmers who cooperate to modify or repair the digital components of their agricultural equipment (Carolan 2018). These micro-scale transformations can then modify macro-scale phenomena, processes that will be detailed in section 2.3.2.

Technological changes have micropolitical effects that are often overlooked (Carolan 2020). Public policies tend to focus on macrosocial effects, such as the number of jobs or the growth rate. However, micropolitical effects, such as impacts on labor, individuals or agricultural practices, must be considered when evaluating public policies. The work of Rotz et al. (2019a) questions the effects of digitalization on salaried workers in the agricultural sector, who could face greater social asymmetry and control. The work of Martin et al. (2021), based on a systematic analysis of the literature on agricultural robots and labor, highlights changes in the nature of the work performed, the workload and the relationship to work – changes that go beyond quantitative shifts. Carolan (2020) shows how the uses of certain digital technologies are associated with a productivist mindset and limit possible futures. The author emphasizes the need for in situ approaches to capture these effects.

2.3.1.4. *Uses of digital technology and transformation of agricultural production models*

The theoretical and empirical literature presented in this chapter raises the question of the microeconomic mechanisms of transformation of agricultural systems through the introduction of digital technologies. The processes of digitalization and ecologization will necessarily interact at the farm level, since the uses of digital technology will be able to modify agricultural decisions and practices. If we revisit Dobrov's concept of technology, which associates it with three components – hardware, software and orgware – digital technology will impact these three aspects. First, digital technology modifies technical and material possibilities. It will enable the performance of new tasks. For example, the precise geolocation (RTK) of a seeder allows sowing under living cover, and digital tools make certain tasks easier, such as milking, field monitoring and so on. On the other hand, digital technology will impact knowledge and skills: data collection tools can create new knowledge, and other tools can externalize and codify knowledge, potentially enabling faster and larger scale information processing, but also potentially diminishing a certain degree of control and decision-making autonomy for farmers. Communication and information tools also provide access to knowledge sources and can generate knowledge through exchange. Finally, digital technology impacts organizational possibilities: farm management, workforce management, adaptability, coordination capabilities and interactions between stakeholders – whether by simplifying, improving, complicating, deteriorating or modifying. These three aspects, hardware, software and orgware, are linked and intertwined, influencing one another. For example, the increased possibility of interaction between agricultural professionals can allow them to lend each other machinery: the material possibilities are thereby modified. The economic and social model of the agricultural enterprise can therefore be affected by technological development; for example, digitalization can lead to an increased commodification of agricultural services or to the outsourcing of certain tasks (Wolf and Wood 1997). Thus, the uses of digital technologies will – in interaction with a set of economic, institutional, social, cultural and individual factors – impact farms and their practices.

Our approach to innovation systems is thus deepened via the use of multiple analytical scales. The study of the microeconomic scale appears complementary to that of the sectoral scale and seems essential for understanding the interactions between digitalization and ecologization. The study of digital uses allows us to understand how digital technologies are used, how they are integrated into agricultural models and how they interact with agricultural practices.

2.3.2. *Interaction between analysis scales*

The different scales of analysis appear complementary for analyzing innovation processes. Beyond being complementary, micro- and meso-scales are interdependent and in constant interaction. Indeed, the innovation system and individual practices mutually influence each other: "The evolution of the sector modifies the exercise of agricultural activity; in turn, the evolution of concrete activity situations can influence the production of standards" (Laurent et al. 2003, p. 146). This section specifies the mechanisms by which the innovation system can impact individual practices as well as those by which individual practices can in turn affect the innovation system, and finally the particular role of intermediary actors in these two-way interactions.

2.3.2.1. *Effects of the innovation system on individual practices*

Changes in individual practices are embedded within the innovation system. The implementation of innovations on a farm depends on a variety of factors. Indeed, a farm is both a microeconomic unit, with production processes, flows, constraints and diverse activities, and a component of a socioeconomic system (Laurent et al. 2003). Decisions made within the farm therefore depend on factors and constraints internal to the farm, whether individual or related to the socioeconomic system in which the farm is embedded (from the egocentric network to the global system), as well as on perceptions and beliefs. Technical change in farming can therefore be seen as a negotiation process between a variety of factors, including techno-economic factors, knowledge, personal objectives (e.g. working conditions or comfort), individual characteristics, work organization, social interactions, public policies and political-institutional determinants (Darré et al. 1989; Ansaloni and Fouilleux 2006). Thus, changes in farms will depend on factors beyond the farm itself: for example, knowledge available in the sector, proposed technologies, regulations in place, public policies and other institutions. However, transformations at the sector level do not reach all farmers in the same way. For example, small farms, or "new entrants" – that is, people from outside the agricultural sector – are less integrated into advisory networks (Mundler et al. 2006; Sutherland 2015). Changes in farms will therefore depend on the forms of integration of farms into the innovation system (Klerkx and Leeuwis 2009a).

The type of integration of farms within the sector will then influence their use of digital technologies. Several studies demonstrate the importance of agricultural extension services in the adoption of digital technologies on farms (Busse et al. 2015; Konrad et al. 2019). Extension services will not only provide knowledge about technologies but also address potential skills gaps in their use (Newton et al. 2020). Eastwood et al. (2012) also highlight the importance of knowledge-sharing networks in the use of digital technologies. Bechtet's (2023) work demonstrates the need for investment to enable extension services to help farmers evaluate agricultural innovations.

The study of innovation systems can thus contribute to a better understanding of the development of digital technology at the farm level.

2.3.2.2. *Effects of individual practices on the innovation system*

Changes in individual farm practices result in the establishment of new routines, which can modify the innovation system. Indeed, changes at the farm level can impact not only demand but also knowledge, innovation creation and institutions. Users are therefore considered full-fledged actors in the concept of a sectoral innovation system. First, the evolution of individual practices will impact demand and thus influence the behavior of organizations in the sector, which can orient their innovation to respond to changes in demand (Malerba 2004). In particular, evolving knowledge needs can give rise to new consulting markets (Laurent et al. 2021). Second, the farm will be a site for knowledge construction (on-farm experimentation, testing of new tools, knowledge of the local ecosystem, etc.), for the circulation of knowledge and learning, but also for the selection, organization and maintenance of knowledge (Cohendet and Llerena 1999). Through interaction and agglomeration effects, the knowledge generated by farms contributes to the knowledge available in the agricultural sector as a whole. Furthermore, changes occurring on farms affect the legitimacy of knowledge and can transform the regulations governing knowledge development by agricultural extension actors (Laurent et al. 2021). Third, beyond simply responding to demand, users can be sources of innovation and can act at different stages of the innovation process (Darré 1996). They can highlight problems, innovate in agronomic practices, create or modify agricultural equipment, and cooperate to develop, test and evaluate innovations. Finally, changes in individual practices can also become institutionalized, transforming uses into social practices. This means that the ways in which a technique is used become recurrent and integrated into a set of cultural practices that can reproduce and that resist forms of change (Badillo and Pélissier

2015). The evolution of individual practices also contributes to the development of new norms of legitimacy and can shift power dynamics (Laurent et al. 2021). The literature recognizes that institutional change is not solely the result of exogenous changes or global institutions, but is also influenced by changes in individual practices (Micelotta et al. 2017). Individual practices thus impact institutions, whether they remain informal or become formalized in norms, rules or laws. Therefore, changes at the micro level can transform the innovation system.

In the field of digital agriculture, there is little literature that addresses this type of phenomenon. The work of Laurent et al. (2021) shows that the development of digital technologies contributes to changing agricultural advisory standards. For example, digital decision-support tools on farms add new forms of advisory overlay and can affect the relative importance of certain types of advice over others. Fielke et al. (2020) also highlight changes in advisory services: digital technologies lead to forms of specialization in decision-making expertise. Furthermore, the development of ICTs has modified the informal advisory network used by farmers, as it has made geographical proximity less important than it once was (Laurent et al. 2021). Digitalization, by offering new data sources and measurement tools, influences environmental policies (Fielke et al. 2019). Digital innovations on farms can influence the labor market and employment relationships, but also the structure of the market – for example, by promoting the concentration of businesses via data control (Rotz et al. 2019b). Thus, individual practices will impact existing institutions. This will potentially affect the coexistence of paradigms within the agricultural sector. The study of individual practices is therefore essential for understanding both institutional and broader changes. However, the literature that delves into the study of individual practices related to digital technology remains limited and rarely examines the overall effects of these evolving practices.

2.3.2.3. *The role of intermediaries in innovation systems*

The literature on innovation highlights the role of intermediaries in the historical dynamics of technological change in agriculture (Klerkx and Leeuwis 2009a). An innovation intermediary can be defined as an organization or entity that establishes a link between two or more parties at a stage of the innovation process (Howells 2006). These intermediaries thus participate in the dissemination of knowledge and technologies, facilitate collaborations and negotiations between actors, clarify innovation needs, mobilize resources, capitalize on experiences and knowledge, and coordinate innovation processes (Faure et al. 2018). Intermediation concerns knowledge, but also infrastructure, regulations and value creation. Intermediaries in agriculture can be of several types. They can be consultants for farms or groups of farmers, digital platforms or databases, frontier organizations between research and users, frontier organizations between research, political actors and training, or even collective organizations (Klerkx and Leeuwis 2009a). The role of intermediaries is being brought to the forefront again with the new challenges of ecologization agriculture, but also of digitalization (Steyaert et al. 2016; Lucas and Gasselin 2018; Newton et al. 2020).

Intermediaries are therefore organizations or actors that facilitate interactions between the micro and meso levels. Embedded in the agricultural sector, intermediaries cannot be entirely impartial; they are active actors in shaping innovation: this is what is called the "paradox of impartiality" (Klerkx and Leeuwis 2009a). Indeed, intermediary organizations establish informal relationships with actors in the innovation system, formalize partnerships with some and also act according to their own interests and the resources they mobilize. Consequently, contradictions can emerge between the actions they implement in this context and their ideal function of neutral intermediation, particularly when the organizations have public and private funding. This makes it all the more important to examine their role in the innovation system.

Collective organizations hold a special place in the agricultural sector, not only as intermediaries, but also beyond. Historically, collective organizations such as trade unions and advisory bodies have played a central role in the modernization of agriculture, and more recently in its ecologization. As intermediaries, these organizations connect sectoral dynamics with individual trajectories. They act as knowledge-intensive service providers; as organizations for experimentation, dissemination and transfer of technologies and knowledge; and as market interfaces; they play a coordinating role between stakeholders and lobby public policy (Howells 2006; Labarthe 2009).

In the context of agricultural digitalization, the role of intermediaries is highlighted but remains understudied. Agricultural advisory services, or the presence of agricultural cooperatives, influence the dissemination and adoption of digital technologies. In some cases, advisors act as catalysts in the adoption of digital innovations by farmers (Bechtet 2023). Beyond this role as "brokers", intermediaries contribute to the construction of meaning and knowledge, and to sharing the problems encountered (Ayre et al. 2019). They participate in forms of mediation that are essential to the success of innovation (Fielke et al. 2020). In particular, they can provide some farmers with the skills needed to use these technologies, thereby helping to adapt them and create value on the farm (Newton et al. 2020). In turn, the introduction of digital technologies can modify intermediation relationships, either by adding a new technological artifact or by introducing new actors. The development of digital services by intermediary organizations can lead to new constraints, but also new sources of revenue (Wolf and Nowak 1995).

To conclude this section, it is therefore essential to emphasize that innovation is both the result and a determinant of development paradigms and models. Produced by the innovation system, innovation will in turn modify and leverage the asymmetries and power relations between actors. "The impacts of innovations thus imply politicizing their implementation with regard to the risks they may entail, the purposes they serve, and the choice of target actors" (Gasselin et al. 2021, p. 94). The system and individual practices mutually influence each other during the innovation processes. Combining these two scales of analysis allows us to study innovation in its various dimensions and in its complexity. This perspective goes beyond a view of farmers as isolated and passive users of sectoral innovations, which they would independently decide to adopt or not. It allows for the integration of factors, constraints and power relations within the sectoral innovation system, as well as processes of institutional change "from below".

2.4. Conclusion

The analysis developed in this work consists of an integrated approach in order to take into account both the structural and dynamic, sectoral and individual components of agricultural transformations. Indeed, the meso- and micro-scales are in constant interaction: transformations in the sector impact the evolution of farms, which can in turn modify sectoral institutions. We study, on the one hand, sectoral changes as both a result and a driver of changes in farms and, on the other hand, changes in farms both as a result and as a possible driver of sectoral changes. To this end, we analyze changes as integrated into a heterogeneous AIS, in which the actors act according to their strategies and perceptions, and interact, in particular via the so-called intermediate organizations of innovation systems.

This theoretical positioning leads us to reformulate the initial research question and to break it down through a set of sub-questions.

NOTE.– How does digitalization interact with the French agricultural innovation system, its paradigms and ultimately with the trajectories of agricultural ecologization? This question can be broken down into three sub-questions.

How do actors in the French agricultural innovation system perceive and implement digitalization, and does this depend on the ecologization paradigm in which the actors are situated?

How do farmers perceive and use digital technologies on their farms, and do their uses depend on their agricultural production model and their ecologization model?

How do agricultural cooperatives, as intermediary actors in the innovation system, integrate digitalization, and how does it interact with ecologization?

The first sub-question relates to the need to understand developments at the innovation system level, which characterizes the actors and networks, institutions and organizations that interact with and impact innovation, and therefore the trajectories of digitalization and ecologization of agriculture. The analysis of innovation systems generally remains limited to a rather static characterization at a given moment. However, the innovations they bring will modify their components, or even create new components or remove some. Conversely, the system and its history will shape the development of innovations. Indeed, the historical trajectory will have an influence through the cumulative nature of progress and knowledge, through established routines, and through all existing institutions. Structural analysis will allow for the identification of actors, associated infrastructures and the mechanisms of interaction between them. This approach also makes it possible to consider the institutions that shape the ways in which these actors act and interact, as well as the technologies and knowledge involved. The dynamics of transformation will be explored via the study of innovation processes at the sector level and their power relations. To conduct this study, we will examine the perceptions and representations of stakeholders, as well as their actions and projects related to digitalization. The stakeholders considered here represent a diversity of organizations within the innovation system and its paradigms.

To fully understand the trajectories of digitalization and ecologization, it is also necessary to understand how companies innovate at the heart of the sector, where transformations are taking place that can, in turn, impact the sector (second sub-question). Thus, our analysis of the digitalization and ecologization of agriculture will also focus on those who implement innovations: farmers. They occupy a special position since they are both stakeholders in the innovation system, actors in its evolution and embedded in – and affected by – it: their practices will be influenced by the innovation system. The evolution of agricultural production models affects the sector, particularly via collective action, the production of standards, knowledge and regulations. Simultaneously, the evolution of the sector impacts individual practices through institutions, knowledge that is created and disseminated, interactions and so on. Moreover, it is the change at the level of farms that is of interest here, which can be considered as the very objective of the analysis of the innovation system under study.

Agricultural cooperatives, since their emergence at the end of the 19th century, have participated in the socioeconomic transformations of agriculture. Our analysis will therefore focus on agricultural cooperatives within the processes of interaction between the digitalization and ecologization of agriculture (third sub-question). Cooperatives, both as a system of interaction between actors and as an intermediary organization, will be examined. Integrated within an AIS, cooperatives play a strategic role in innovation processes. They are places where decisions are made, knowledge is transferred and created, training and information are provided, economic activities are carried out, and the interests of cooperative members are expressed and defended. Through these various functions, cooperatives contribute to interactions between the micro- and meso-scales and can influence innovation processes.

3

Digital Technology and Ecological Agriculture: Multi-scale Transformations

In order to present the interactions between digital development and the ecological transformations of agriculture, the remainder of this work will focus on three scales of analysis associated with three fields and three sets of results. This multi-scale approach, which stems from the theoretical issues presented in Chapter 2, thus allows us to observe and analyze digital development "as it unfolds", both in practices and in organizational and institutional transformations (Schnebelin et al. 2022a).

A first scale of analysis is that of the agricultural innovation system, within which agricultural organizations, businesses and administrations interact and evolve via the joint transformation of knowledge, technologies and institutions. The second scale of analysis is that of farms, within which farmers use digital technologies, develop their practices and innovate, while adapting to institutional and economic transformations. The third scale of analysis is that of cooperatives, intermediary organizations embedded in the agricultural innovation system and in markets, which coordinate a range of activities for producers. These three interdependent and co-evolving scales are schematically represented in Figure 3.1.

In this chapter, we detail the methodologies developed at these three scales, specifying how the theoretical framework presented in Chapter 2 is operationalized, what data we use and how we analyze it.

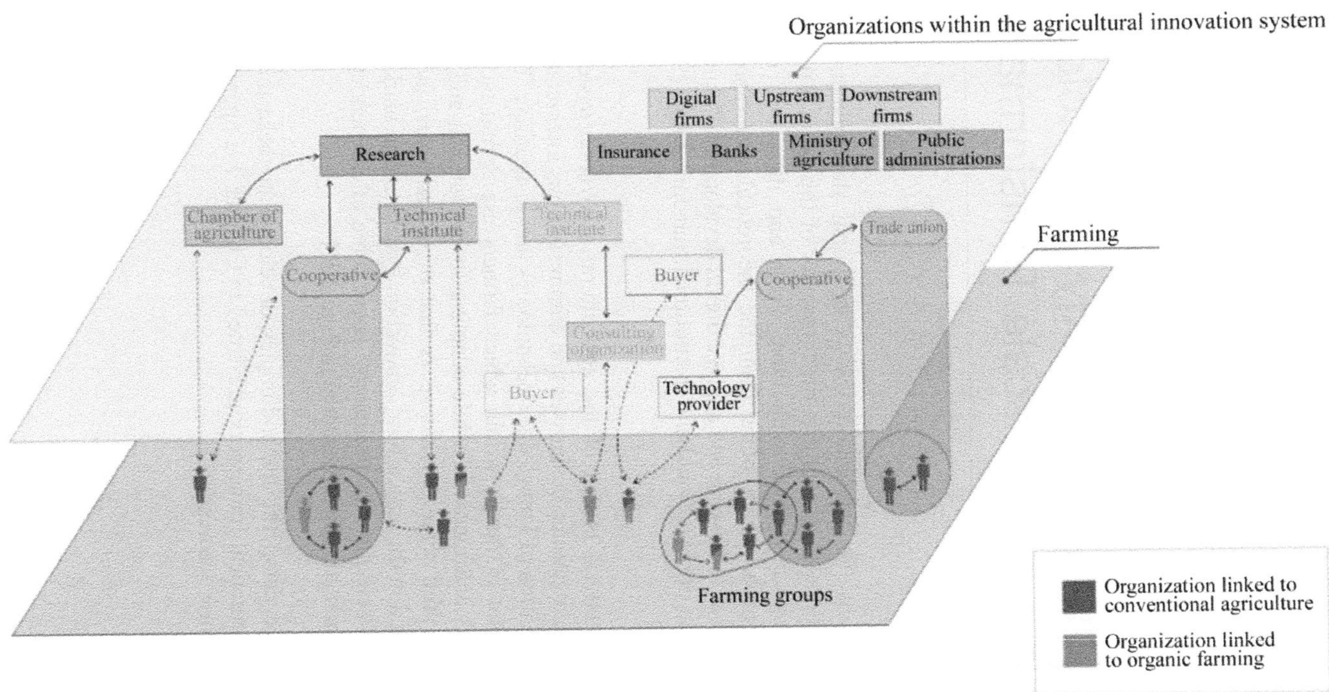

Figure 3.1. *Visualization of the multi-scale approach of the book (this diagram was created with the assistance of Delphine Jacquot, Atelier Khroma). For a color version of this figure, see www.iste.co.uk/schnebelin/ecologization.zip*

FIELD OF STUDY.–

For our field investigations, we are interested in one country in particular, France, and especially the Occitanie region.

Since the end of the Second World War, a genuine agricultural innovation system has taken shape in France, marked by the emergence of education and research institutions, agricultural advisory organizations as well as a set of institutions supporting investment at the farm level (access to credit, price support, subsidies, tax measures, etc.) (Gerbaux and Muller 1984; Servolin 1989). This innovation system, combined with a set of regulations, has enabled the modernization of agriculture and an increase in agricultural production, associated in particular with an increase in the average farm size and capital intensity, as well as a decrease in the agricultural population. Today, this innovation system is evolving with the arrival of digital technologies, with the emergence of projects and networks aimed at promoting research on the subject, along with a set of public policies supporting digital technology in agriculture (support for innovation and investment). Together, these developments make France a leading country in the growth of digital agriculture, while integrating it into environmental considerations (Bellon-Maurel et al. 2023). This structuring of an innovation ecosystem around digital agriculture is particularly dynamic in the Occitanie region (see Figure 3.2), where research projects and networks such as DigitAg[1] and OccitaNum[2] have been established, and where a number of companies in the field are developing, such as Naïo Technologies[3], ITK[4] and TerraNIS[5].

Beyond being interesting study areas for digital development, France – and particularly the Occitanie region – offers fertile ground for the study of ecological agriculture. This is due, first, to their location within the European Union, where agricultural policies have been undergoing an ecologization process since the 1990s (Deverre and Marie 2008). Second, a set of national and regional policies structures this ecologization process in France and Occitanie, with the development of several forms of ecological agriculture such as organic farming, integrated pest management (IPM, which aims to reduce pesticide use through the implementation of preventative practices), conservation agriculture (which aims to protect or enhance soil health), etc. (Lamine 2011; Plumecocq et al. 2018). The

1 See: https://www.hdigitag.fr/.

2 See: https://occitanum.fr/.

3 See: https://www.naio-technologies.com/.

4 See: https://www.itk.fr/.

5 See: https://www.terranis.fr/.

Occitanie region in France is a pioneer in these dynamics, with almost a quarter of the country's organic farms (InterBio Occitanie figures 2024).

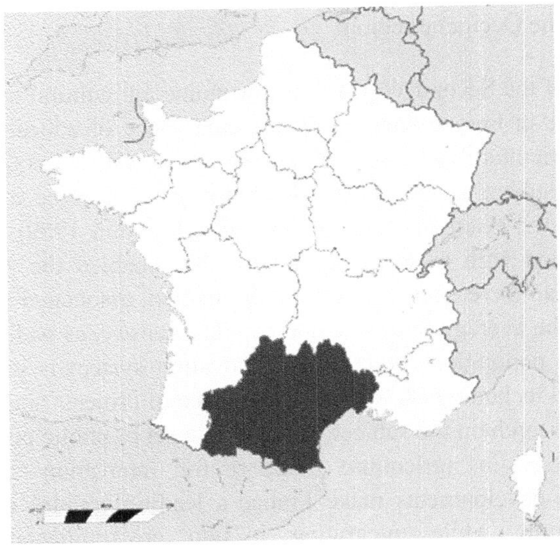

Figure 3.2. *Location of the Occitanie region in France (source: Wikipedia). For a color version of this figure, see www.iste.co.uk/schnebelin/ecologization.zip*

3.1. Perceptions and commitments of actors in the innovation system

In the first instance, we examine the impact of digital technology on the heterogeneous agricultural innovation system. This system can be analyzed through the perceptions and actions of its stakeholders, which can influence individual practices. The objective of this analytical framework is to understand the perceptions and actions of stakeholders within the French agricultural innovation system in relation to the digitalization process, and whether these perceptions and actions vary according to different paradigms. The results of this analysis are presented in Chapter 4.

3.1.1. *Components of the agricultural innovation system*

To analyze digitalization at the sectoral level, we used Malerba's 2002 analytical framework. This framework allows us to integrate the diversity of actors in the sector, its heterogeneity and its dynamics. It therefore appears relevant for studying the interactions between digitalization and ecologization within the agricultural sector.

The concept of sectoral system of innovation (SSI) proposes to study the innovation of a sector through five central components, which will interact and impact innovation: actors and networks, knowledge, institutions, technologies and demand (Malerba 2004).

The agricultural innovation system framework highlights the importance of considering the diversity of actors involved in agricultural innovation. We included traditional agricultural actors, as well as a number of actors specializing in digital agriculture, characteristic of the dynamic "boundaries" of an innovation system (Malerba 2002). We then assembled a purposive sample (Etikan 2016). Traditional agricultural innovation system organizations include research, training and advisory organizations; professional organizations; and collect and supply organizations (cooperatives and trading companies) (Spielman and Birner 2008; Klerkx et al. 2012). Within these organizations, we targeted individuals identified as experts on digital matters or those having an overall view of this topic within their organization. Observations, gray and scientific literature, and data from interviews conducted (snowball sampling technique) made it possible to identify the "new" actors (Atkinson and Flint 2001). This allowed us to identify digital companies expanding their activities into the agricultural sector, companies being created specifically targeting digital agriculture, and existing organizations creating new projects dedicated to digital agriculture. For the farmers included in this study, we selected individuals with representative positions (particularly union representatives). During interviews with farmers, several of them had a role in training and disseminating digital technologies; they were therefore included in the "digital" group.

Thus, this step will consist of a cross-analysis between structural components of the agricultural innovation system and between categories of actors: actors of the conventional paradigm, actors of the organic farming paradigm as well as common sectoral actors and actors specializing in the digitalization of agriculture. This first step of the overall approach can be visualized in Figure 3.3.

3.1.1.1. *Integration of heterogeneity*

The French agricultural innovation system is not homogeneous. Intra-sectoral heterogeneity is significant in the dynamics of sectoral systems of innovation (Malerba 2002, p. 262). Particularly with regard to ecologization, several approaches are proposed and implemented in agriculture. To capture this heterogeneity, we focus here on the dominant paradigm as well as an alternative paradigm: organic farming. Organic farming is one of the most institutionalized examples of alternative agricultural practices, especially in France (Bellon and Penvern 2014). Indeed, organic farming is studied, developed, represented and promoted within a variety of specific organizations: research and science; advisory services; professional organizations (unions and interprofessional bodies); collect, storage and processing organizations; and marketing and distribution organizations (Piriou 2002).

Organic agriculture arose from social and ideological struggles against the development of productivist agriculture. The recognition of organic agriculture within the AIS, which was also framed by and for conventional agriculture, was one dimension of the confrontation between organic and conventional agriculture (Brechet and Schieb-Bienfait 2006).

In France, the first organic farming group was created in 1959, followed by a working group on organic agriculture in 1959, and then by the French Organic Association[6] in 1962. This movement led to the institutionalization of organic agriculture with the creation of the Research Group on Organic Agriculture[7] in 1978, the official recognition of organic agriculture in 1980, and then the creation of the Technical Institute for Organic Agriculture (1982) and the organic label (1985) (Piriou 2002). Thus, the development of organic agriculture is characterized not only by different practices and values at the level of individual farmers and consumers but also by specific institutions and organizations that frame the development of organic agriculture within the agricultural sector. Today, in France, organic agriculture is the most "institutionalized" alternative paradigm. Its growth rate has been over 15% over the last 15 years. Since 2018, organic farmers have supplied more than 6% of French food and account for more than 8% of agricultural land (Agence Bio 2020).

Box 3.1. *The institutionalization of organic farming in France*

This institutional distinction allows us to establish a boundary between two paradigms, a boundary that is relatively easy to define. Organic farming is seen as one form of agroecology, or as a prototype of agroecology, even though agroecology goes beyond what organic farming is (Van der Ploeg et al. 2019). Other forms of alternative agriculture are, however, becoming institutionalized, for example, through certification systems or producer associations.

6 L'Association française d'agriculture biologique, in French.

7 Groupement de recherche sur l'agriculture biologique, in French.

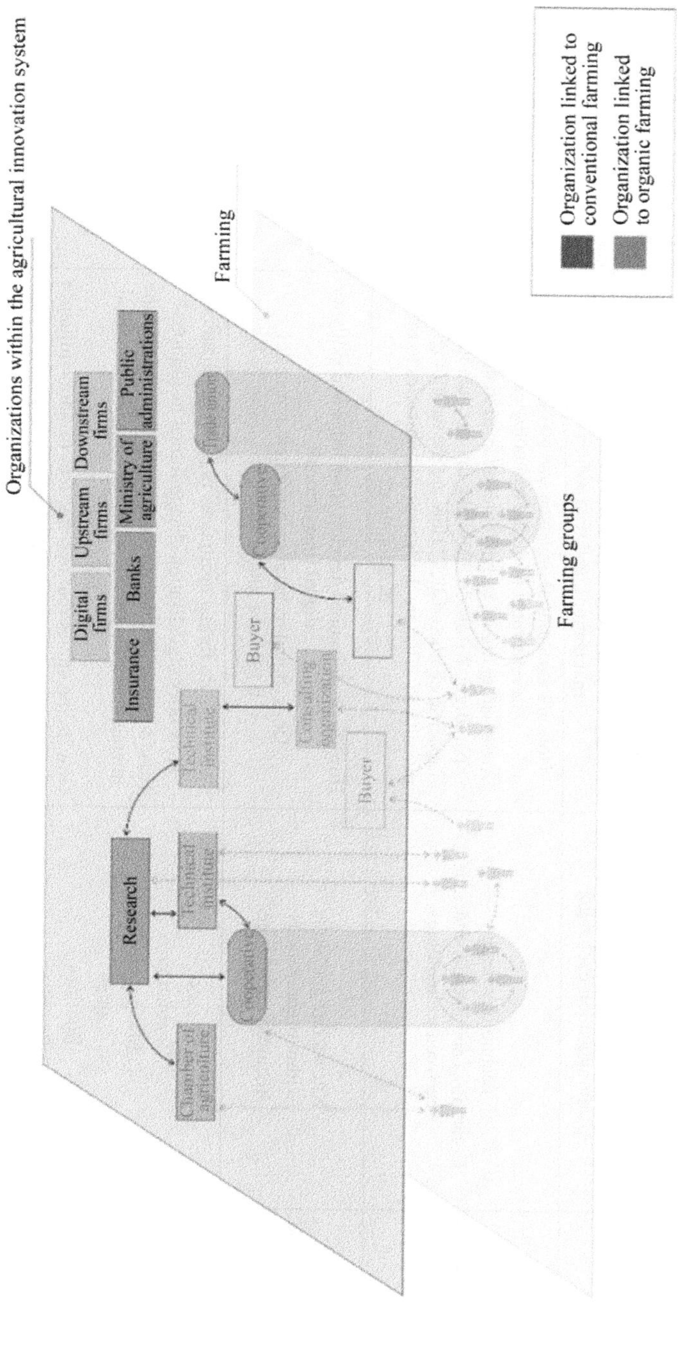

Figure 3.3. *Visualization of the first scale of analysis. For a color version of this figure, see www.iste.co.uk/schnebelin/ecologization.zip*

Category of Malerba	Description	Questions	Literature
Actors and networks	Beliefs, assumptions, objectives Organizations Collaboration/Competition Interactions Communication/exchanges	What do stakeholders expect from digitalization? What risks do they perceive? How does digitalization change interactions between or within organizations? Does it lead to collaboration or competition? Do digital actors work with certain AIS organizations, and if so, which ones?	(Bronson and Knezevic 2016; Jakku et al. 2016; Eastwood 2017; Dufva and Dufva 2018; Rijswijk et al. 2019)
Technologies	Technology design Constraints and interdependencies between technologies	How are agricultural organizations involved in the development of digital technologies? Are existing digital technologies adapted to different paradigms? How do these take into account their specific characteristics? How are the technologies perceived? What are the factors and obstacles to their development?	(Jakku and Thorburn 2010; Carbonell 2016; Bronson 2019; Rijswick 2019; Lioutas and Charatsari 2020)
Knowledge	Knowledge and skills within organizations Acquired knowledge	How do organizations develop knowledge and skills for digital innovations? Does digital technology bring new sources of knowledge ?	(Jakku and Thorburn 2010; Eastwood et al. 2012, 2019; Rijswijk et al. 2019; Ingram and Maye 2020)
Institutions and public policies	Laws Regulation Public policies Values Routines Practices	What role do formal institutions play in digitalization? How does digitalization change formal institutions? How do digital-related institutions take into account the interactions between digital technology and ecological agriculture? How do the values/routines/practices of actors influence digitalization within paradigms and vice versa?	(Wolf and Buttel 1996; Eastwood et al. 2012; Jakku et al. 2016; Wolfert et al. 2017; Rijswijk et al. 2019)

Table 3.1. *Construction of coding systems for interview analysis*

3.1.2. *Focusing on the French agricultural sector*

In order to capture these representations and actions, we conducted semi-structured interviews with actors in the agricultural sector.

The interviews were conducted at the national level across the French agricultural sector. The national level, coupled with the sectoral level, allows for a certain institutional coherence and makes it possible to include the network leaders of professional agricultural organizations in the surveyed organizations. Indeed, many agricultural organizations – unions, chambers of agriculture and advisory bodies – are structured with regional and/or sub-regional entities, as well as national entities, which assume the role of political representation, defending the interests of their organizations and promoting them in public debate. At the level of actors closer to the field – cooperatives and farmers – we remained focused on the Occitanie region, which will be our scale of analysis for the next stage and characterized by the coexistence of organic and conventional farming.

The interviews were primarily conducted between March and November 2019. The interviews were conducted using a semi-structured approach, rather than a standardized questionnaire, allowing for adaptation to the interviewees. Generally, the stakeholders in the innovation system were interviewed with regard to four main areas. The first section aimed to gather information about the organization, its history, its functions and the interviewee's role within it. The second section focused on the organization's activities related to digital agriculture. The third section explored their knowledge of digital technology use in agricultural operations. The fourth section aimed to discuss, more broadly, their perceptions of the potential, risks and links to the ecologization of agriculture. The interviews were recorded and transcribed.

3.1.3. *Abductive coding analysis*

The analysis was conducted in several stages. Initially, the interview notes and transcripts were reviewed without coding (Dumez 2013). Subsequently, the transcripts were coded using MaxQDA® software. This coding was partly based on Malerba's categories (knowledge, technologies, actors and institutions) and partly on the inductively generated sub-codes shown in Table 3.1. The results were first described by actor and category and summarized in a table cross-referencing actors and codes. A second level of analysis was then performed to identify key elements, using an abductive approach – that is, an iterative process between theoretical models and empirical results (Peirce 1974; Lawson 1989). Indeed, the abductive process appears to be relevant from an institutional economics perspective (Labrousse 2006). We then distinguished five elements: the diversity of expectations regarding the development of digital technology, partly linked to the conventional or organic paradigm, the central role of knowledge and technologies, the diversity of digitalization strategies, the positioning of digital actors in relation to the heterogeneity of the innovation system and the risks perceived by actors within these paradigms. We then highlighted the divergences and convergences between paradigms in the discussion.

3.2. Uses of digital technology and the transformations of practices in agricultural operations

In a second step, we analyze the concrete uses of digital technology on farms. The objective of this level of analysis is to examine the links between digital technology use and farm management models in order to understand the potential interactions between digital technology and the ecologization of practices. The results of this analysis will be presented in Chapter 5.

3.2.1. *Investigating practices and uses*

3.2.1.1. *An agricultural production model embedded in an innovation system*

Farming can be conceptualized as a productive agricultural model. This perspective allows us to consider production (type of product, quality, quantity, target markets, etc.), the resources mobilized, whether bio-technical (land, inputs, equipment, etc.) or human (labor, training, farmers' social capital), but also the organization of activities and work in time and space (combination of technical operations or activities, crop rotations, supply, marketing, etc.) and the farmer's project as well as the economic context in which the farms are embedded (market, innovation system, networks) (Bourgeois and Sébillotte 1978). To analyze the farms, we therefore established a questionnaire covering these various structural, individual, socioeconomic, agronomic and technical dimensions. It includes questions about the operator and their background, production, marketing, work (internal and subcontracting), agricultural machinery, agricultural practices, the economics of the farm, their integration into the innovation system via the operator's egocentric network (advisory links, membership in professional groups), as well as the uses and perceptions of digital technology.

3.2.1.2. *Digital usage profiles*

We focused on the actual uses of digital technologies on farms. For each technology, we then incorporated questions about the reasons for use, the concrete functioning, the learning achieved for the use of these technologies, the associated changes in practices and organization, and the advantages and limitations.

To take into account the interdependencies between digital technologies, we have constructed profiles of digital usage rather than studying the technologies one by one.

We focused on two categories of technologies: digital technologies for production (DTP)[8] and digital technologies for information and communication (DTIC)[9]. Here is a brief description of the most common technologies:

– The use of a geolocation system on agricultural machinery allows for guiding (guidance) or automating (auto-guidance) the operation of a machine within a field. The standard accuracy achieved with GPS using differential correction (DGPS) is of approximately 20 cm. Accuracy can be increased to about 2 cm with the RTK system, which requires a paid subscription.

– Section control technology, based on GPS positioning, is a device that automatically closes certain sections of the sprayer, seeder or fertilizer spreader, in order to avoid overlaps in the application of inputs.

– A connected weather station is a set of sensors (depending on the case: rain gauge, hygrometer, anemometer, barometer, thermometer, spectrometer, soil probe) that measure, transmit in real time and store meteorological data. They can be networked (e.g. the different weather stations of a cooperative), which gives each farmer with a station access to the data of the other stations. They can also be associated with decision support tools (DSTs).

– The yield monitor is a sensor which, combined with GPS, allows for the measurement and mapping of yield (and also moisture) of a crop during harvest. The resulting maps show areas where the yield is lower or higher than average. They can potentially be used with modulation technology.

– Variable rate application maps are maps created to characterize intra-field heterogeneities based on soil, vegetation or yield analyses. The recommended doses of fertilizer or seeds are adjusted to these heterogeneities. These maps can be used with a spreader or seeder adapted for variable rate application. In this case, the spreader or seeder adjusts the dose of fertilizer or seed applied as it progresses in order to adapt the dose to the area according to the data from the variable rate technology.

– Various digital DSTs for treatments are offered to farmers. These DSTs are based on disease prediction models and weather data (incorporating or not incorporating data from weather stations). They indicate, for a given plot and variety, whether or not the disease is likely to develop.

---

8 *Technologies numériques de production* (TNP), in French.

9 *Technologies numériques d'information et de communication* (TNC), in French.

– Connected irrigation technologies bring together technologies that allow remote management of irrigation systems, verification of irrigation operation, or access to data on irrigation needs.

– Land management software is software that can be used from a computer, and sometimes from a smartphone, which allows users to record and access farm data, mapping, interventions carried out or recommendations made by an advisor. It also serves as an interface to many digital services (notably DSTs).

For information and communication, we used more generic digital technologies (websites, forums and social networks). During the surveys, we discussed the use of digital technology and the internet for information gathering, communication and promotion on a variety of topics: agronomic and technical information, economic or regulatory information, agricultural equipment, buying/selling products or services, sales and communication. Information-gathering practices were explored in depth: frequency of use, types of websites consulted and participation in or use of social networks (agricultural forums, Facebook pages, YouTube channels, etc.). We also discussed the impact of this information on farming practices.

3.2.1.3. *Characterizing ecological agriculture*

To account for the diversity of organic farming practices, we included questions on labels and certifications, marketing channels, and agronomic practices. In collaboration with agronomists, we identified a set of agronomic practices related to input management, soil protection and cultivated biodiversity, representative of a range of practices that vary in their degree of ecological sustainability. Soil protection can be assessed through variables related to tillage type and soil cover (establishment of cover crops). Cultivated biodiversity can be understood via interspecies diversity (legumes, cereals, oilseeds and grasslands) or intraspecies diversity (varieties), as well as spatial and temporal diversity (intercropping, crop rotation and cover crops). Input management encompasses fertilization management (mineral, organic and crop rotation), plant health management (pesticides, biocontrol, varieties and crop rotations) and weed management (herbicides, mechanical weeding and crop rotations). For these three areas of ecologization, we have developed clusters to categorize farm ecological practices.

3.2.2. *Focus on the field crop sector in Occitanie*

To study these innovation trajectories, we are examining a specific sector: field crops. First, this sector is emblematic of French agricultural modernization and is the subject of extensive research and technological development, particularly in the digital field (Lowenberg-DeBoer and Erickson 2019). This sector is perceived as economically strategic in France and has significant investment capacity. However, this sector also raises important environmental issues due to its substantial consumption of synthetic fertilizers and pesticides, leading to water, air and soil pollution. Various proposals for ecologization practices in field crops have emerged and become institutionalized, with the significant growth of organic farming, as well as soil–conservation techniques and biocontrol, along with initiatives to diversify and lengthen crop rotations, and even agroforestry.

To obtain a sample encompassing a wide diversity of farming models and practices, we chose the Occitanie region as our study area. On the one hand, it is a region where large-scale farming plays a significant role, covering more than 1 million ha. On the other hand, this region is a leader in France in the digital development for agriculture, notably with the establishment of the DigitAg convergence institute and the presence of numerous companies in this field. Furthermore, it is also a very dynamic region in terms of ecologization, with projects and support initiatives focused on this theme, and significant growth in organic farming. Occitanie is the leading region in France for organic field crops, accounting for 24.5% of French production (Interbio Occitanie 2018).

To analyze the complexities of uses and practices, we conducted face-to-face interviews with 98 farmers. The interview guide, developed collaboratively by an interdisciplinary team, included a set of closed-ended questions to systematically characterize farm models. These closed-ended questions covered farm structure, individual characteristics of the respondent, production and practices, marketing and the socioeconomic network. Technical datasheets were used to explore digital technology usage in more depth. A more open and general discussion on the issues surrounding digital technology in agriculture was conducted at the end of the interview.

3.2.3. *Mixed method of analysis*

This data informed mixed-methods analyses. Closed-ended questions and the creation of a database from these interviews enabled quantitative analyses of the data. The first quantitative analysis method used is classification analysis or clustering. This involves grouping individuals into relatively homogeneous categories, called clusters, based on a set of specified variables. By categorizing respondents in this way, using variables that characterize their digital technology usage, classification analysis helps to identify distinct profiles of digital technology use. A second quantitative analysis consists of statistical tests to compare means or percentages between these groups. This method allows for the calculation of means (for quantitative variables) or percentages (for qualitative variables) for variables in different groups, and for testing whether there are significant differences between these groups. This allows us, for example, to see if there are differences in farm size between individuals with different usage profiles. These quantitative analyses highlight links and correlations between variables. To complement this quantitative analysis, we conducted a qualitative analysis based on transcripts of interviews with ideal types for each profile. The qualitative analysis allows us to study the respondents' perceptions, see their representations, understand their decision-making processes, grasp their uses in detail, understand temporal dynamics on their farms and thus formulate explanatory hypotheses about the links between variables. It also allows us to understand the role of the different actors in the agricultural innovation system at the farm level. Interviews with farmers holding representative roles in the agricultural sector were also included in the analysis presented in section 3.1. The responses to the more open-ended interview questions were subjected to textual and qualitative analyses. These were used to produce the interview reports at the farm level and for further analyses.

3.3. Intermediate organizations: between diffusion and transformation of innovations

3.3.1. *Intermediaries and cooperatives in innovation systems*

3.3.1.1. *Innovation intermediaries*

An innovation intermediary is defined as:

> An organization or body that acts an agent or broker in any aspect of the innovation process between two or more parties. Such intermediary activities include: helping to provide information about potential collaborators; brokering a transaction between two or more parties; acting as a mediator, or go-between, bodies or organizations that are already collaborating; and helping find advice, funding and support for the innovation outcomes of such collaborations. (Howells 2006, p. 720).

This "intermediary" role can be central or secondary to other organizational functions (Klerkx and Leeuwis 2009a). These authors identify three main components of the intermediation function: (1) identifying and articulating innovation needs; (2) to form networks and facilitate links between actors; (3) to manage innovation processes (disseminate knowledge and technologies, guide research, mobilize resources, create legitimacy and avoid resistance to change).

The importance of these intermediary organizations for innovation has been highlighted, particularly in agriculture (Klerkx and Leeuwis 2008). They appear to play, or could play, a central role in addressing the new challenges of ecologization

(Kivimaa et al. 2019; van Lente et al. 2020) but also of digitalization (Eastwood et al. 2012; Newton et al. 2020) in the context of the privatization of these organizations (Labarthe 2009). This raises the question of how the role of these intermediaries is transformed by digitalization and ecologization, and how these two processes are articulated by these organizations.

3.3.1.2. *Cooperatives as key intermediaries in the French agricultural innovation system*

The International Cooperative Alliance defines a cooperative as "autonomous association of persons united voluntarily to meet their common economic, social and cultural needs and aspirations through a jointly owned and democratically-controlled enterprise" (ICA 1995). From the 19th century onwards in Europe, agricultural cooperatives developed, enabling farmers to improve their bargaining power, bypass certain intermediaries, access markets, products or services, reduce costs through economies of scale, control quality, and also manage and pool risks (Chiffoleau et al. 2008; Valiorgue et al. 2020). Initially, the ethical principles of self-organization and mutual assistance were central, aimed in particular at combating the economic, social and political domination of large-scale merchants or farmers (large wine estates, for example). Cooperatives, their roles and their raison d'être have evolved, accompanying modernization and mass production, crisis management, internationalization and diversification. Gradually, they have assumed a political role of co-management with the State and a technical role, notably through the implementation of R&D, consulting and training.

Today, cooperatives are seen as central organizations in providing advisory services to farmers (Laurent et al. 2021). Yet, they are rarely studied as intermediaries for innovation. The work of Yang et al. (2014) shows that, through their position as collective organizations, they generate contextualized and integrated knowledge, adapted to the complex realities of farms and practices. Through their multifunctionality, they combine the technical, economic and social dimensions of innovations. However, as economic and political organizations, they are also criticized for the potential existence of conflicts of interest, particularly between input sales and advisory functions, which do not necessarily seek to reduce the use of inputs for ecological purposes (Chomel 2019). While representing the interests of (the majority of) their members, they would not, however, be "neutral" intermediaries with regard to sectoral issues (Yang et al. 2014).

Beyond their intermediary role, cooperatives are unique in that they are collective organizations founded on democratic principles and owned by member farmers who are also their suppliers (collect and marketing cooperatives), users (supply and service cooperatives) or both (multipurpose cooperatives). The relationships between member farmers and their cooperative must therefore be analyzed as a complex and original system of interactions between actors. They are therefore characterized by governance mechanisms (Saïsset 2021) and activity coordination systems, and are framed by legislation and laws specific to agricultural cooperatives, but also by less formal social interactions between member farmers (who are often neighbors) and between them and other actors in the territories (Touzard and Draperi 2003; Chiffoleau 2005; Saïsset 2021).

As both an institution and a system of interaction between actors, the cooperative is a dynamic organization that is constantly evolving. This evolution results from both internal and external transformations within the cooperative. Thus, following Cook's 2018 analysis, cooperatives follow established cycles:

1) an emergence phase, where actors converge their interests, develop a project and decide on the justification for cooperation;

2) then an organizational design phase where the functioning of the cooperative is defined concretely and adjusted;

3) then followed by a phase of growth, during which heterogeneities and divergences of interest would appear, creating frictions, coordination problems related to different production portfolios, "free rider" situations or varying activity horizons, differentiating the expected returns from the cooperative's investments and prompting some members to leave the organization;

4) the cooperative would then enter a phase of analyzing frictions and resolving these coordination problems, leading to several possible outcomes: the crisis could lead to a halt, merger or absorption by another organization, maintenance through adjustment of the project and internal rules, or even a more global restructuring that could go as far as a modification of the organization's statutes (e.g. the weighting of votes, the creation of a subsidiary, or, in the United States, the creation of a "new cooperative" with, for example, the opening of capital to non-cooperators).

Digitalization and ecologization can intervene in these cycles by modifying the justifications for cooperation and especially the heterogeneities between members, the differences in capacity, projects and interests with regard to these two processes.

When it comes to digitalization and ecologization, the question arises of how cooperatives' "raison d'être" will evolve, and how will their consulting activities, internal transformations and their capacity to address the new "frictions" and problems posed by the differentiations associated with these two processes affect their members? Indeed, digitalization and ecologization can modify both the system of intermediary actors within the innovation system, impacting knowledge-sharing and advice mechanisms, and the internal governance structures of cooperatives (Duvaleix-Tréguer et al. 2019; Fielke et al. 2020).

3.3.2. *Dual representation of agricultural cooperatives*

In order to analyze the transformations in the roles and positioning of cooperative wineries with the processes of digitalization and ecologization, we have therefore chosen to study the cooperative both as an institution of an innovation system, and as a system of interaction between actors. Our study is therefore based on a dual and dynamic representation of the wine cooperative, an economic organization considered both as a system of interactions between actors and as an institution inserted into a sector and a territory, but also into an innovation system.

The wine cooperative can first be considered as a system of interactions between actors that implements and coordinates activities fulfilling a variety of functions, designed to benefit its members. These activities are organized around the harvesting of grapes and winemaking, followed by the bottling and marketing of the wines. These operations rely on collective investments, are partly carried out by cooperative employees, are coordinated by rules and agreements, and are part of a strategy defined by a board of directors and approved at a general assembly (Saïsset 2021). Member winegrowers are thus engaged in contractual relationships, support collective decisions and must resolve a series of coordination problems related in particular to the heterogeneity of their situations (Bareille et al. 2017; Cook 2018). But beyond these formal and statutory relationships, the cooperative winery is driven by other social interactions, linked to internal committees, projects or events and more broadly to networks of dialogue and advice among members and between them and other local stakeholders (Chiffoleau 2005). Through all these networks, the cooperative contributes to the dissemination and sharing of knowledge. It can therefore guide the technical choices of its members both through its investments, its strategy and its collectively developed rules, particularly the rules for classifying and paying for grapes (Touzard 2010), but also through the various networks of dialogue that drive it and contribute to its governance.

The wine cooperative is also a legally recognized institution, a social economy company embedded in the wine sector. It develops economic relationships within this sector to market its wines, purchase goods and services, take out loans and so on. The economic justification for cooperative wineries is linked in particular to the market power they can establish in a sector where markets are highly segmented and regulated (Geographical Indications, traceability, etc.). Cooperatives also participate in political relationships to defend their interests and influence wine policies (via Coop de France, for example) or to support investment projects that require political backing. Within the sector, they are also at the heart of information flows and advisory relationships with dedicated organizations (chambers of agriculture, interprofessional organizations, the French Wine and Vine Institute [IFV][10],

10 Institut français de la vigne et du vin, in French.

oenologists, the media, etc.), as well as with other cooperatives within clusters (Chiffoleau and Touzard 2014). The cooperative winery can then be seen as a component of a wine innovation system (Boyer and Touzard 2017), in which it plays the role of an innovative actor, for example, for its winemaking or marketing activities, but also an intermediary role (Klerkx et al. 2010) with regard to the innovations implemented by its members. It can indeed promote and guide winegrowers' access to economic resources and information that will influence changes in practices on their farms.

These two complementary representations of cooperative wineries allow for a dynamic analysis: the interactions of the cooperative's actors, which extend well beyond the organization itself, can constitute "innovative actor systems" (Chiffoleau et al. 2008) that influence the organization's strategy and performance, its evolution and its positioning within the sector. In turn, the cooperative institution imposes itself on its members as a framework for action and an interface with the market and the entire innovation system, influencing the evolution of their practices. This approach has proven its relevance for studying the innovations of the "quality revolution" in Languedoc wineries since the 1980s, highlighting the importance of intangible resources and the capacity to foster "responsible business" projects (Chiffoleau et al. 2008), but also demonstrating the challenge of resolving a series of coordination problems, redefining their governance and trajectory (Cook 2018; Saïsset 2021).

We will use this dual and dynamic vision of cooperative wineries to analyze how they participate in ecological and digital transitions. We therefore adopt an organizational approach that considers the expectations and perceptions of its stakeholders, the interests that bind them to the organization and also their interactions, which contribute to collective decisions, innovation and organizational transformations within an innovation system.

3.3.3. *Focusing on the wine industry in Occitanie*

The wine sector thus appeared as a relevant area for studying the role of cooperatives as collective and intermediary organizations within the agricultural innovation system. The decision was made to remain within the same administrative area: the Occitanie region. On the one hand, this allowed for the study of a different sector within a similar territorial institutional context. On the other hand, the Occitanie region boasts the largest vineyard area in France (approximately 270,000 ha) (Chambre d'agriculture Occitanie 2021). Furthermore, it is the leading French region in terms of organic vineyard area (34,827 ha, representing 37% of French vineyard area in 2018). The wines produced cover a diverse range of markets (appellations, PGI, domestic market and export).

To select the sites, we proceeded in several stages. First, we spoke with a manager in the wine sector at Coop de France Occitanie. This exchange provided an opportunity to discuss the challenges and questions facing the industry, and to identify cooperatives with diverse profiles in terms of digital development and their commitment to environmentally friendly projects and production methods (organic, sustainable viticulture, High Environmental Value[11] [HVE], etc.). Our contact at Coop de France provided us with a list of 18 cooperative wineries illustrating this diversity, indicating their geographical area, production area and volume, number of members and some comments on the cooperative's projects (certifications, European projects, use of digital technologies). We then contacted these cooperatives, aiming for a diverse sample in terms of size, structure and type of production. Eight cooperatives were ultimately selected (the sample is described in the article). The method is based on a series of interviews conducted in 2020 with managers of these wine cooperatives (directors or vineyard managers) as well as winegrowers who are members of these cooperatives (between 4 and 10 members per cooperative, 43 in total). In addition, 11 interviews were conducted in vineyards geographically close to these cooperatives to gain a better understanding of the context, but these were not included in the analysis.

As with field-crop farmers, we sought to meet a diverse range of profiles among winegrowers belonging to the surveyed cooperatives, both in terms of production methods and digital technology use. The selection was made following interviews with cooperative managers, which may have introduced a bias (members generally aligned with cooperative management), even though we consistently strove to cover a variety of situations, including members close to the board of directors as well as others less so.

The interviews with the managers of cooperative wineries focused on (1) the organization of the cooperative (size, production, marketing and governance), (2) involvement in projects related to agroecology, (3) activities related to digital technology in the cooperative, and (4) the uses of digital technology by winegrowers and the perceptions of the people surveyed.

At the winegrower level, we adapted the questionnaire developed for field crop farmers, expanding the questions on their relationship with the cooperative. The data from the interviews were entered into a database. This allowed us to inform their strategies for ecologization and digitalization at their level and to link these strategies to those of the cooperative wineries. We also identified the winegrowers' perceptions of their cooperative's role in both ecologization and digitalization projects.

---

11 *Haute valeur environmentale*, in French.

3.3.4. *Cooperative strategies*

The analysis of these data was conducted in several stages, following a qualitative method involving categorization, cross-referencing information within each cooperative and paying close attention to the justifications and explanations provided by the interviewees. Initially, for each cooperative, we identified every project related to ecologization or digitalization. For each project, we recorded the start date, partners, motivations, obstacles and limitations, operational details, and perceived advantages and disadvantages. Subsequently, a comprehensive analysis of the cooperatives' projects allowed us to identify the areas and objectives of action for ecologization or digitalization. This enabled us to situate these processes within the various strategies of the cooperatives. To understand these strategies, we identified the actors involved, explained the rationales for action and the functions of the cooperative wineries. We identified the role of the cooperative and its transformations both as a commercial interface and as a coordinating organization between actors.

3.4. Conclusion

In this work, we therefore propose a multi-scale and systemic analysis of agricultural transformations. Within this framework, our analysis will begin with the actions and representations of actors at the various scales considered in order to analyze these transformations beyond general representations, examining their microeconomic and political effects. Part 2 of this work will present the results obtained from these three approaches. A conclusion will synthesize, integrate and articulate the results from the three areas, stages and chapters in a comprehensive analysis of the digitalization and ecologization of agriculture.

PART 2

The Development of Digital Technology in Agriculture at Different Scales

4

The Development of Digital Technology in the Agricultural Innovation System

This chapter examines the relationship between digitalization and ecologization trajectories at the innovation system level, that is, at the level of institutions, organizations, networks and actors that interact with and impact innovation in a given space (Touzard et al. 2014). This chapter aims to understand how actors embedded in different agricultural paradigms, in this case organic and conventional farming, perceive and implement digitalization. Institutions, knowledge and technologies at the sector level are modified by and for digitalization and can influence the development of digital tools and their uses on farms. However, the heterogeneity of the agricultural sector means that these changes do not necessarily impact all actors and farmers in the same way.

Before examining the differences in terms of uses on farms, we will focus here on the differences in perceptions among actors in the agricultural innovation system and their involvement in the digitalization process. Based on semi-structured interviews with actors in extension services, research, agricultural cooperatives, technical institutes, unions and businesses, we analyze the divergences and convergences between organic and conventional agriculture in the digitalization process, both in terms of objectives, perceived challenges and risks, and the strategies implemented. We also analyze how this heterogeneity is perceived and taken into account by actors specializing in the development of digital agriculture. This chapter is based in particular on work published in July 2021 in the *Journal of Rural Studies,* written with Jean-Marc Touzard and Pierre Labarthe (Schnebelin et al. 2021).

The first two chapters have shown that the development of digital technology in agricultural operations is not merely an individual matter, but involves broader changes in the agricultural sector. Questioning the compatibility of the digitalization of agriculture with its ecologization therefore requires analyzing not only farmers' practices and uses, but also sectoral developments.

This chapter provides an overview of the digitalization of agriculture, while examining its compatibility with ecologization from the perspective of actors within the innovation system. Specifically, this chapter addresses the following questions: What are the perceptions and actions of actors in the agricultural innovation system regarding digitalization? Do these differ depending on whether the actors work primarily in conventional or organic farming? Do digital actors collaborate with organizations in the agricultural sector? If so, with which organizations? How do actors in digital agriculture take the diversity of forms of agricultural ecologization into account?

As explained in Chapter 3, we will focus here on the case of France. In France, the agricultural innovation system is structured around co-management between the State and the agricultural sector, which notably shapes R&D, training and advisory services (Allaire and Boyer 1995).

Since the 2010s, the theme of digital technology for agriculture has gained momentum within the French agricultural innovation system, with research projects, training programs and the arrival of new actors originating from the digital world, innovation funding policies or even the organization of professional events on the theme of digital technology in agriculture (Martin and Schnebelin 2024).

4.1. Perceptions of actors in the agricultural innovation system

The different categories of actors – those belonging to digital agriculture organizations and those representing conventional and organic paradigms – perceive and implement digitalization differently. Table 4.1 summarizes the actors' statements according to the different categories used for data analysis. The following sections present the results according to the five major issues that emerged: the diversity of expectations, the key role of knowledge and technologies, the new interactions between actors generated by this cross-sectoral transformation, the specific role of digital actors in the agricultural innovation system and the crucial issue of perceived risks.

4.1.1. *Diversity of expectations according to organic and conventional paradigms*

The actors mentioned various expectations regarding digitalization (see the "Overall vision" column in Table 4.1). Some impacts of digitalization are anticipated by all. These include the optimization of practices, access to information and advice, traceability, risk management, and improved technical and economic management of farms. Farmers also mentioned convenience and time savings.

However, differences are also observed regarding certain expectations concerning communication with consumers, knowledge and value creation. Conventional actors view digitalization more as a means of creating new economic opportunities, while organic farming actors see it more as a means of developing knowledge. A number of opportunities identified by those interviewed relate to communication with consumers.

Conventional stakeholders primarily mentioned traceability as a way to improve communication and marketing of agricultural products. One interviewee quoted a statement heard during a meeting with a representative from a major retailer: "We sell a product, that's true, but what we we're missing is the product's story". Thanks to digital technologies, organizations can ensure increasingly precise traceability and can hope to gain added value.

Organic farmers are increasingly viewing digital technologies as a means to improve sales, deepen interactions with consumers or facilitate direct sales.

The development of environmental regulations and private standards (such as the implementation of the HVE 3 certification in wine, or CRC4 in cereals) promotes digitalization tools that align with traceability.

> The regulatory obligation to record practices, manage organic fertilization, register with the Common Agricultural Policy, etc., is what really pushed farmers towards digitalization. (Agricultural advisor/conventional advisor)

Another set of opportunities relates to the emergence of a new market based on data and digital technologies. Some conventional agricultural organizations consider embracing digitalization and being able to offer digital services to their farmers to be a sound business strategy. They are investing in digital technologies to ensure their continued presence in the advisory market and to find "new business models" within the current legislative framework (particularly the requirement to separate sales and advisory services).

For some of these organizations, the objective is clear: to sell services. Furthermore, digital technologies are considered essential for addressing the challenges facing agriculture: environmental impacts, animal welfare, profitability, working conditions and attractiveness.

Digital technology is considered "the future of agriculture" and as a prerequisite for their future survival. It is also seen as a way to improve the image of the agricultural world in the eyes of society by participating in conveying the image of a modern sector that takes environmental issues into account.

> We therefore have a market [that of plant protection products] which will probably decline. We must therefore position ourselves in relation to other niches that can be sources of profit. (Trade/Conventional)

> This will help farmers show society [...] that they are doing better and better and that they are ready to take advantage of all new technologies to improve their production. (Union/Convention)

Following the same logic of digitalization, stakeholders in agricultural digital technology emphasize the potential advantages of digital technology: gains in productivity, yield, time, security, forecasting, better management and communication, simplification, and efficiency. Data are considered as a value, as "intangible capital" (AgTech Company 2).

For these stakeholders, digitalization is considered essential for the future of agriculture in order to address agricultural challenges, including environmental issues, climate change and new societal expectations. They evoke a necessary and inevitable transformation that will revolutionize agriculture.

The use of digital tools in agricultural practices is considered intrinsically good and sustainable, so digitalization becomes an end rather than a means. This development of digital technology "is very politically supported" (digital agriculture researcher) and is heavily backed by funders and research.

Members of organic organizations add expectations regarding learning and assistance with system design, assistance with systemic farm management, link building, knowledge exchange, sharing experiences and the opportunity to make better observations.

	Knowledge		Technologies		Actors		Institutions	
	Capabilities	Creation/ exchange	Development	Constraints	Global vision	Interactions	Formal	Informal
Organic	The importance of developing farmers' skills. Lack of projects at the interface of digital technology and organic farming.	Digital technology allows us to capitalize on and exchange knowledge.	Developing technology in-house to facilitate knowledge capitalization and exchange.	Many technologies are not suitable for technical, organizational, or economic reasons.	Potential for managing complexity and systemic vision. Risk of dependence, loss of know-how and power.	Few partnerships with digital players: a different vision of digital technology – some informal exchanges.	Environmental standards are driving digitalization – there is no public support for technology to cooperate.	A vision of agriculture sometimes opposed to digital technology, because it is based on autonomy and reduced costs/ investments.
Conventional	The need to develop skills within agricultural organizations to support digitalization.	The need to manage data to create value within agricultural organizations.	Adoption of new technologies, co-development.	Issues surrounding data ownership. Misuse of technology by farmers.	Potential for renewing the economic models of agricultural organizations, improving the image of agriculture, improving efficiency – risks to data ownership.	Collaborations to test, promote, or co-develop tools. Risks of market lock-in.	Regulation as a driving force behind digitalization – the need for regulations to protect farmers (personal data, interoperability).	Farmers' habits seen as a major obstacle to digitalization.

Specializing in digital technology	Farmers' skills = obstacle to digitalization. Organizations specializing in digital technology can contribute their skills to analyze agricultural data.	Data and digital technology will help agriculture to experiment, model, predict, etc.	Technology needs for digitizing agriculture: Digital technology adapted to all forms of agriculture, including organic farming.	Problems with data access, quality, compatibility and complexity. The problem of a suitable business model.	Digitalization is in its early stages and will be necessary to address environmental and economic challenges. Data as intangible capital.	Agricultural organizations need to reach farmers. Data sharing is necessary. Governance issues are at stake.	Regulations underpin digitalization, but hinder the development of certain tools.	Farmers' habits seen as a major obstacle to digitalization.
Transverse	Regulations are driving digitalization. Need to train farmers in digital technologies.	Potential for creating knowledge with digital technology, but also a need for cooperation, sharing and resources.	No development of technologies.	Potential of digital technology for environmental sustainability; issue of adapting to a diversity of agricultural systems.	Potential for implementing public policies. Risks of unforeseen effects.	Digital technology is a factor in increased exchanges between actors in the agricultural sector.	Regulation is seen as a factor in digitalization, but the private sector remains a driving force in innovation.	The agricultural sector needs to renew itself to enable radical innovations.

Table 4.1. *Summary of interviews according to the categories of actors interviewed*

4.1.2. *Knowledge and technologies at the heart of the digitalization process*

Beyond these expectations and promises regarding digitalization, the interrelationships between knowledge and technology were highlighted as major challenges by all actors. A need for knowledge emerges with digitalization, while digitalization generates opportunities for the creation of new knowledge. First, a consensus emerged on the need to acquire new knowledge and skills to embrace digitalization. Respondents in conventional agriculture emphasized knowledge at the organizational level, while those in organic farming focused on knowledge at the farm level (see the "capabilities" column in Table 4.1). Conventional agricultural organizations, such as agricultural cooperatives, stressed the importance of developing new types of knowledge within their structures. Specialized jobs and teams are thus being created specifically for digitalization, and awareness-raising and training activities are being conducted. Internal positions in agricultural organizations are sometimes even filled by digital specialists.

> Farmers increasingly need experts [...]. This forces us to train ourselves differently, or even to train people on certain aspects, etc. (Cooperative 2/Conventional)

Organic farming organizations are placing increasing emphasis on the need to develop farmers' skills. Those interviewed agree on the necessity for farmers to acquire new knowledge in order to increase their autonomy and enable them to master the basic tools for managing their farms.

> And computer literacy is essential for us [...] so that people can be self-reliant. We don't think it's complicated, but [some say] it's too complicated for farmers and that it's not their area of expertise. We say it's possible to use the basic tools, and this creates critical thinking about their farms. (Consultant/Organic)

Developing skills at other levels, such as research and development, has also been mentioned by stakeholders in organic farming, for example, by the French Scientific Committee for Organic Agriculture[1]. However, these organizations have limited resources and other priorities.

The stakeholders agree that the development and use of new technology creates new data that could contribute to the development of new information and knowledge. The second column of Table 4.1 summarizes the respondents' statements and shows that stakeholders in organic agriculture emphasize the creation

1 Le Comité scientifique français de l'agriculture biologique, in French.

of knowledge about agronomic practices, while conventional agriculture actors emphasize the creation of information through traceability.

According to stakeholders in organic farming, digital technologies in organic farming would be useful for obtaining information on regulations, trade and machinery; for analyzing and understanding ecological processes; for helping farmers design or develop their own systems; for analyzing their practices; and for allowing farmers to consider their own specific characteristics and choices. The capitalization and sharing of knowledge appear to be a key advantage of digitalization, and these stakeholders mentioned a "conversion support tool" or a "design support tool" to help farmers transition to organic farming. They primarily believe that digitalization can provide new "knowledge inputs" for designing, evaluating and sharing their farming practices. This will, however, still require physical and concrete approaches. Digital knowledge exchange is seen as a way to complement real-world exchanges, but not replace them.

Conventional actors are placing greater emphasis on generating information through traceability technologies in order to "better meet the standards of the value chain and strengthen consumer trust and knowledge about products". Traceability is increasingly demanded by buyers (i.e. major distributors, wholesalers and exporters), but it is difficult to implement. Collective organizations hope to create knowledge through data collection. However, they struggle to process their data due to a lack of resources.

Thanks to digital technologies, digital companies hope to create new knowledge that will drive the development of their own business: digitalization could create new forms of experimentation, new tools to carry out comprehensive analyses of agricultural practices and environmental criteria, to improve modeling and forecasting.

> We are convinced that over time, a lot of know-how will come out of the vineyard. We are at the very beginning of the process, because the accumulation rate is not very high, so it takes time. (AgTech Company 1)

4.1.3. *Different partnership strategies with digital actors*

Intersectoral dynamics are seen as a major factor driving the digitalization of the sector. Digitalization brings new actors and partnerships into the agricultural sector. Startups and companies from other sectors are investing in agriculture, leading to new types of interactions between actors (see the "Interactions" column in Table 4.1). One might think this limits the role of agricultural organizations, but this is not the case. Agricultural organizations – cooperatives, associations, chambers of agriculture, trading companies and advisory service providers – play a central role, particularly in data collection, but also in data redistribution and technology dissemination. Many digital actors claim they cannot access farmers directly. They need intermediaries based on farms to collect the large quantity and diversity of data required for data-driven tools to function. Agricultural organizations are also needed to legitimize digital projects.

> The goal [for our company] is not to sell directly to farmers, but to sell to cooperatives or traders or management centers – who will be the distributors of our solutions to farmers, because they have an interest in collecting and federating data to carry out their work [...]. (AgTech Company 2)

However, we noted differences between the paradigms. Digitalization is perceived by conventional actors as an exogenous change and by organic actors as a more endogenous one. Conventional organizations work in partnership with digital actors at various levels: to test, co-develop or promote digital tools. These interactions can be informal or formal. When agricultural organizations collaborate with a digital company, they position themselves as distributors, but also as service providers. They also offer support and training to farmers. In other words, they want to transform technology into a service they can sell to farmers. Conventional organizations consider partnerships with digital companies as strategic. Digital technologies are seen as one element of a service, associated with inputs, advice, machinery and knowledge, all coordinated by data management. According to one of the interviewees, this could lead to a market lock-up and reinforces their view that digitalization is an important business strategy for them.

Organic farming stakeholders are less involved in collaborative projects with new digital players. On the one hand, digital players do not often call upon or work with actors from organizations specific to organic farming.

> But by working with everyone in a balanced way, we mostly work with those who are most visible. And we don't work much with small producers, with agroecology. (Digital agriculture researcher 1)

On the other hand, when organic organizations are approached, things do not always go smoothly due to differences in working methods and values. Furthermore, organic organizations have other priorities and lack the financial resources to invest further in digitalization.

> Each time, the choice, the cultural difference is a little too strong. Even if we have a similar attitude towards environmental issues, our methods are very different. (Advisor/Organic 3)

Although the digitalization of organic farming is thus envisaged in a more endogenous way, the organizations have informal exchanges with digital actors and are interested in and follow the development of digital technologies.

Digital technology developers consider the development of partnerships between organizations to be strategic. They argue that digitalization will require a different organizational structure for the sector, an ecosystem for developing information systems. Sharing data and ensuring compatibility are essential for achieving effective digitalization. Beyond strategic partnerships, some digital stakeholders regret the limited role given to farmers in digital projects.

4.1.4. *Low awareness of the internal heterogeneity of the agricultural innovation system by digital actors*

Digital actors bring a new perspective to the innovation system. They highlight governance problems between different categories of actors, but do not perceive differences between organic and conventional farming.

Digital stakeholders claim to want to support agriculture through the digitalization process. Digitalization is considered an end in itself for the agricultural sector, which will need to digitize to improve its economic and environmental performance. According to these stakeholders, farmers are not aware of the benefits of digitalization and are not particularly attracted to the idea of using digital technologies. The digital organizations interviewed either directly develop technologies (startups, established companies), are involved in technology development projects (research, companies) or test technologies (training organizations). Digital companies working in other sectors want to transfer their technologies from other sectors to the agricultural sector.

> We need to evangelize, to make people understand the ins and outs of what we do. (AgTech Start-up)

Digital organizations believe that digital technologies are equally suitable for organic and conventional farming. They do not consider organic farming as a differentiating factor.

> In fact, at least since the beginning of the project, I don't get the impression that being organic or not influences the interest people

> have in it. I feel like it's a cross-cutting issue. (Digital Agriculture Training)

Digital organizations see several obstacles to their development in the agricultural sector. First, they mention several barriers regarding data access, including data quality, technological compatibility and interoperability, the cost of technologies and constraints related to the specific context of agriculture, notably long-term timeframes, variability and complexity. Second, they highlight governance issues concerning data management. Third, regarding the acceptability of their technologies, they are aware that digital technologies lead to outsourcing some of the analysis, which can discourage farmers from adopting them. Fourth, they emphasize potential problems related to farmers' ability to pay for and use digital technology.

> To work with Big Data and analysis, you need good quality data. And that's difficult to obtain. (AgTech Business Association)

> And in all projects, regardless of the technology, the weak link is governance. (AgTech Business Association)

> When we use an interface like ours, there's this idea that [farmers] are outsourcing part of the data analysis and that they have to accept it. And I think that's very difficult to accept. (AgTech Company 1)

4.1.5. *Major difference between paradigms: the risks perceived by actors*

All actors highlighted risks associated with the opportunities they mentioned. Organic farming stakeholders emphasized the risks related to knowledge, while conventional farming stakeholders emphasized the value of data. Organic organizations and farmers listed numerous risks: in particular, the potential for these technologies to be too expensive, the risk of becoming dependent on them and losing power, the "risk of standardization" and the risk of data hacking or misappropriation.

Other risks were mentioned, such as stress or time commitment, loss of face-to-face interaction, loss of connection to the land and loss of local knowledge. Specific problems were mentioned when farmers lack the necessary digital tools or the skills required to use them.

Digitalization sometimes clashes with the philosophy of some organic farmers or is simply too disconnected from their way of life. In particular, organic farming can reduce costs and investments, while digital technologies may require investment. Not all digital technologies currently under development are considered suitable for organic farming, whether for technical or socioeconomic reasons: they may not fit the economic model, the farmers' ways of thinking and making decisions, and so on.

As one farmer pointed out, he cannot properly use his agricultural software because it is not designed for a comprehensive approach to his farm: it is designed for a specific technical itinerary or field management rather than for general management. The reservations expressed by farmers are reflected in a comment made by an advisor:

> But in our approach to advising, we believe that in organic farming, decisions must truly take into account the entire farm [...]. It's necessary either to visit the farm or at least speak on the phone and provide truly personalized advice. (Consultant/Organic 1)

Digital technologies are complex, and their complete mastery does not seem possible for these actors. This could alter the balance of power between actors.

> Beyond the loss of know-how, it is the balance of power in an agricultural system that will be disrupted. In other words, we will be very dependent on the equipment or services provided within the framework of these systems, and on data processing, which is sometimes a bit of a black box as well. (Advisor/Organic 3)

To ensure that technologies meet their own requirements, organic organizations may develop them in-house, often through a bottom-up innovation process: an innovation is designed, implemented and tested locally, and then, if it works, it is scaled up. Most of the technologies developed by organic farming organizations concern knowledge management and exchange.

> So we obtained the tool at the national level, we invested money to use it and improve it based on the feedback we had already received, and that was a good thing because we had a very good foundation. (Union/ Organic)

Conventional actors highlighted the risks associated with data ownership, particularly the risk that digital service providers would capture all the value created. They also mentioned the risk of exclusion for some farmers due to a lack of infrastructure and skills or due to cost. Farmers mentioned additional risks regarding the reliability of digital technologies and dependence on repairs, and highlighted the risk associated with the additional cost of equipment when farmers are already facing economic problems.

Uncertainty regarding the value of data, farmers' ability to understand the potential of technologies and the misuse of tools by farmers are cited as constraints by organizations involved in the development of digital technology. For their part, farmers have emphasized the need to better integrate real-world conditions into the design of digital technologies.

It is therefore clear that various visions of digitalization coexist. Depending on the vision of digitalization they convey, the institutions that frame digitalization could thus influence the direction of this trajectory.

4.2. Digitalization beyond paradigms

Our research confirms that digitalization not only changes technological possibilities, but that it participates in the reorganization of the entire agricultural innovation system in interrelation with multiple factors, as suggested by previous studies (Busse et al. 2015; Fielke et al. 2019; Rijswijk et al. 2019). The interactions between actors, knowledge and institutions are jointly modified by digitalization within the agricultural innovation system, revealing characteristics common to different ecologization paradigms.

Regardless of their paradigm, agricultural organizations play a significant role in digitalization, acting as intermediaries between digital companies and farmers, but also as proactive actors in digital development and in the collection, analysis and transfer of information. Digitalization does not diminish the role of intermediaries but can even strengthen it, as Busse et al. (2015) demonstrate. This is yet another illustration of the role of "innovation brokers" in agriculture (Klerkx and Leeuwis 2009b).

All the actors interviewed agreed on the potential of digitalization to improve working conditions, optimize practices and manage risks. They also mentioned the possible advantages for the economic management of farms, traceability, consumer information, and information and training for farmers. The digitalization of agriculture thus falls within the framework of "technoscientific promises" (Joly 2015).

Furthermore, all those interviewed mentioned different potential risks that could limit the adoption of digital technology or lead to the exclusion of farmers. Economic risks for farmers are described as being related to the cost of technologies, a lack of skills or dependence on outsiders to repair machinery. With the exception of "digital farmers", farmers agreed on other risks concerning data hacking or appropriation by value chain actors. They also mentioned the risk that technologies might not be suitable for small farms. These findings are consistent with perceptions of digitalization in the New Zealand AKIS and of big data in the Australian grain industry (Jakku et al. 2016; Rijswijk et al. 2019).

The need to maintain some control over the ongoing digitalization process was mentioned in both paradigms. Actors in the agricultural sector want to be proactive agents of digitalization rather than passive recipients. They aim to reach the final stage of *digi-grasping* described by Fielke et al. (2021). All interviewees emphasized that digital innovations must be integrated with other types of innovation, not just technological ones. Rotz et al. (2019a) acknowledge that this is a major challenge for digitalization.

Digitalization affects knowledge in a reciprocal process: it creates a need for new knowledge about digital technology while simultaneously generating new knowledge itself. The creation and dissemination of knowledge constitute a major evolution, as demonstrated by Fielke et al.'s (2020) literature review. However, making this knowledge effective proves complicated due to the diversity of needs and contexts, and the management of complexity, among other factors. Several organizations claim to possess data but are unable to analyze it due to a lack of resources. Various cross-sectoral stakeholders even believe that the value of data is a myth: they believe that agricultural actors hope to exploit the value of data, which will not happen.

Regulations, standards and specifications were considered by those interviewed as major drivers of the accelerated development of digital technologies. Digital technology can be both the cause and the consequence of evolving regulations, enabling the establishment of new types of regulations and new forms of control and traceability (Pearson et al. 2019).

On all these points, digitalization appears more as a source of convergence than divergence between actors in relation to the conventional vs organic paradigm. This convergence results from the perception of shared advantages (better information, easier work, etc.) or problems (autonomy, learning and the evaluation of technologies, etc.). These analyses thus constitute a basis for reflection or action for a digitalization that integrates the diversity of agricultural systems.

Convergence can also be linked to the evolution of the paradigmatic dichotomy, this distinction having become less clear (Sonnino and Marsden 2006). The rapid development of organic farming leads to hybridization mechanisms between organic and conventional organizations. On the one hand, "conventional" organizations, particularly cooperatives, are extending their activities to organic agriculture

(Stassart and Jamar 2009). On the other hand, organic farming organizations are integrating innovations that allow them to scale up and "become conventional" (Le Velly and Dufeu 2016).

The distinction between the two paradigms and their institutions remains relevant. However, in practice, a kind of continuum is increasingly observed. Thus, some "conventionalized" organic actors may have a "conventional" view of digitalization.

4.3. Diversity of digitalization trajectories envisioned

Although this digital transformation is global, it is not perceived in the same way by all actors, and points of divergence exist between organic and conventional actors regarding their digital appropriation of this transformation (Dufva and Dufva 2018). Digitalization could therefore reinforce the differences between paradigms and trajectories of the agricultural innovation system.

The main differences between actors in organic agriculture and those in conventional farming seem to lie in the direction each expects from digitalization:

– Digitalization in the service of traceability is expected by conventional actors, while organic actors raise the risk of standardization, fearing that the "industrialization" of organic products will result from standards linked to or imposed by digital technologies aimed at promoting traceability (Ringsberg 2014; Klerkx et al. 2019; Rotz et al. 2019a).

– Digitalization to create endogenous knowledge is expected by organic stakeholders, who hope that digital technologies will help them design and analyze their production systems systematically and support experimentation. However, this is not how digital technologies are currently designed; they are more segmented than holistic, more top-down than bottom-up. This could lead to divergences between digital technologies and organic farming. Organic agriculture stakeholders mention the potential risks of loss of power and know-how.

– Digitalization for value creation is expected by conventional and digital stakeholders who hope to improve the image and attractiveness of agriculture, increase profitability and limit environmental impacts. Conventional farmers and their organizations raise concerns about data ownership.

Beyond envisioned directions, we observe that different innovation processes and digitalization strategies are being implemented. Organic farming stakeholders emphasize the importance of farmer training and the design of specific technologies to support their own vision of digitalization. Conventional stakeholders collaborate with digital actors to make farmers' activities simpler and more efficient. Thus, the actors involved in digitalization differ because organic farming organizations focus on internal development, while conventional organizations develop technical and economic partnerships.

However, during our interviews, the digital stakeholders did not perceive these different viewpoints. They work with the most influential players and see no difference between organic and conventional farming. They consider most digital technologies to be generic and therefore suitable for both conventional and organic farmers.

However, the knowledge base differs between organic agriculture and conventional agriculture, and for success, farmers' knowledge must be incorporated into digital technologies (Rose et al. 2018). Involving stakeholders in the design of tools is essential for end users to understand them (Jakku and Thorburn 2010; Bronson 2019).

Failing to consider diversity within innovation systems, and consequently failing to integrate this diversity into tool design, could lead to the exclusion of non-conventional forms of agriculture. This could reinforce the dominant paradigm. Conversely, a diversity of digitalization could in turn strengthen their diversity.

Here, we consider organic farming as an example of an agroecological transition paradigm in France. It constitutes a prototype of ecologically oriented agriculture (Bellon and Penvern 2014), but other paradigms exist, and organic farming itself is not homogeneous and encompasses a diversity of practices and representations. Furthermore, the diversity of conceptions of digitalization depends on a variety of factors, not just paradigms. This opens up research opportunities to study digitalization for other forms of alternative agriculture, in other locations, or according to other factors.

4.4. Analysis of innovation systems reinforced by taking into account heterogeneities and power relations

The structural analysis based on Malerba's framework highlighted the transformation of the agricultural innovation system for and by digitalization, while taking into account changes in the very nature of the system's components, intersectoral dynamics and its heterogeneity. This analysis allowed us to highlight both convergence and divergence within the innovation system regarding the digitalization process in agriculture.

Our findings are consistent with Fielke's (2019) results, which showed that digitalization leads to power issues and highlighted that historically powerful operators can appropriate more gains from digitalization. Therefore, power issues can arise between different actors (AgTech actors versus farmers, for example). We also add potential power issues between different types of agricultural systems and different visions of digitalization to this. Research by Bronson (2019; Bronson and Knezevic 2016) confirms that digital technologies are primarily geared toward conventional agriculture. Our research is complementary in providing insight into how digitalization could be useful to organic farming. It appears that there is no opposition to digitalization itself, but rather to a certain definition of digitalization that currently prevails. This conception of digitalization tends to be prescriptive, requiring significant investment, concentrating power and standardizing production. It is supported not only by private actors, but also by certain public actors (Lajoie-O'Malley et al. 2020).

This situation calls for the inclusion of the concept of paradigm and power relations within the innovation system. It invites researchers to analyze not only how digitalization occurs, but also its potential direction and how it is envisioned and supported by the innovation system. Cross-sectoral actors could work with digital stakeholders to raise awareness of this issue and promote diversity in research and development to avoid potential lock-ins related to digitalization. This raises the question of digitalization governance. Governance will influence the opportunities that digitalization will address, the risks it will avoid, and, consequently, the agricultural paradigm it will promote. In line with the conclusion of Newton et al. (2020), it is essential to involve farmers and citizens in decisions concerning the digitalization trajectory. We add the need to involve a diversity of agricultural systems to promote their diversity. In this regard, functional and relational analysis could complement this work by identifying barriers and incentives (Bergek et al. 2008). The directionality of change also depends on the use of digital technologies by producers and the constraints they face, which in turn calls for further research on the concrete uses and practices of digital technology by farmers.

Our analysis encourages us to take a step back when referring to the concept of digitalization. In practice, digitalization is not a single phenomenon with a single definition: it does not mean the same thing to different actors. Digitalization can have different objectives, occur in different ways and take different forms. We show here that there are no different "stages" or "levels" of digitalization depending on the actors. All actors are engaged in understanding, becoming aware of and transforming digitalization. But we suggest that there are different "processes" of digitalization. However, this raises questions about the possibility of the coexistence of different digitalization processes, or whether power imbalances will impose a standardized digitalization, meaning that only the future imagined today by dominant actors will become a reality (Carolan 2020). Our findings therefore call for the inclusion of heterogeneity in innovation systems to enable the development of technologies adapted to different ecologization trajectories. We provide conceptual and empirical elements to help stakeholders become aware of this heterogeneity. Furthermore, many interviewees emphasized that digital technologies are only one component of the transformation; the others include changes in advisory services, farm structure, new relationships with consumers and new policies supporting open innovation. Thus, the success of digitalization should not overshadow other dimensions of agricultural systems, and it is necessary to further explore their interrelationships.

4.5. Contributions and perspectives

This analysis at the level of the innovation system raises several questions and perspectives.

The observation of these differences between paradigms at the innovation system level raises questions at the farm level. Are these oppositions also found at the individual level? Are the uses of digital technologies on farms linked to different agricultural models and the ways in which farmers are integrated into the innovation system? Are digital uses associated with certain practices of ecologization agriculture rather than others? The heterogeneities of expectations and perceived risks, highlighted in this work, invite us to ask whether such heterogeneities exist at the farm level. Our results call for studying the farm level to also see if other mechanisms of differentiation are at work there, revealing *bottom-up* transformations of innovation systems (Laurent et al. 2021). These questions appear all the more crucial given that the results show a lack of knowledge among stakeholders in the agricultural sector regarding farmers' use of digital technologies. They will be addressed in the next chapter.

Furthermore, the results demonstrate the dependence of digital organizations on intermediaries. The development of digital technology on farms can be linked to the intermediary organizations with which they interact. Understanding the role of these intermediaries for farms is key to analyzing their integration into the innovation system and the evolving relationships between paradigms. Farms that are not connected to these intermediaries, that are outside these networks, may therefore lack access to certain technologies. For digital organizations, the farmers involved are those integrated into the networks. This may lead to reflecting on potential problems of excluding certain farmers in digitalization, especially since the literature shows that not all farmers are linked to intermediaries, particularly those who have small farms or those who do not come from the agricultural sector (Labarthe and Laurent 2013; Sutherland et al. 2017; Kinsella 2018).

The results of this chapter also raise questions about the positioning of these intermediary organizations in their various functions. They can play a technological role, such as intervening in technological innovation processes, participating in design or testing and experimentation, disseminating technologies and supporting or training farmers. They also have an organizational role that can be modified, such as their interactions with actors in the agricultural sector, with digital agricultural companies, and with farmers, as well as their internal functioning and organization. Digitalization can also transform their role as economic intermediaries: the choice of economic models for digitalization, investments, outsourcing and partnerships. Finally, it raises questions about their political positioning, and in particular the governance mechanisms that have been put in place in this development. These questions will be addressed in Chapter 6.

5

The Development of Digital Technology in Agricultural Operations

At the level of organizations within the French agricultural innovation system, digitalization is perceived and implemented in heterogeneous ways. But what about at the level of individual farms? This chapter aims to examine the concrete links between digitalization and ecologization at the farm scale. Adopting and implementing new technologies on a farm involves changes in organizational modes, knowledge management and agricultural practices. To fully grasp these changes, it is necessary to go beyond approaches that focus solely on "adoption" by questioning how technological innovations are concretely integrated into the farm system, leading to adjustments and learning processes. It is therefore essential to analyze how digital technologies are actually used and how these uses are constructed. However, little is known about digital uses at the farm level. The main objective of this chapter is to capture digital practices at the scale of farms, as well as the logics and conditions associated with these uses. Although these practices are developed by individuals (farmers or farm workers), they are considered to be embedded within a broader socioeconomic system and may thus be influenced by institutions, value chains, egocentric networks and the training of these actors, as well as by the farm's production model. To study these uses and their conditions in relation to production models and the innovation system, we focused on farmers specializing in arable crop production in the Occitanie region, conducting 98 interviews. This substantial fieldwork is central to the analysis presented in this book. This empirical investment allowed us to move beyond a vision of digitalization shaped by institutional discourses, commercial communications or, more broadly, technoscientific promises. It is, among other things, a matter of going to see everyday digitalization in its practices, its learning and its bricolage (Higgins et al. 2017).

The survey conducted with 98 farmers make it possible to study which digital technologies are used, how, why and what they change in practices. With a sufficiently large sample of farms, the integration of data on digital uses, farm structures and agricultural practices gives the possibility of statistically analyzing the links between the agricultural farm model and the use of digital technologies.

FIELD OF STUDY.–

The uses of digital technology in agricultural operations are explored here in the field crop sector in the Occitanie region. This region is an interesting case because it is the site of (1) both a strong dynamism in terms of agricultural digital technology, with the establishment of numerous companies and research projects, and (2) both a significant and multifaceted development of ecological agriculture, with a particularly substantial share of organic farming in this region.

HYPOTHESIS CONSTRUCTION.–

To analyze the relationship between digitalization and ecologization in agriculture, we focus on two types of digital technologies, which potentially have different impacts on ecologization trajectories.

Digital technologies for production (DTPs) aim to improve the efficiency of agricultural production processes. They were designed with the goal of optimizing the conventional agricultural production system. Some authors argue that these technologies only make sense within such systems. Consequently, we can imagine that economic variables such as utilized agricultural area (UAA), economic size and labor will have positive and significant effects on the adoption of DTPs. Furthermore, outsourcing certain tasks to an agricultural contracting company (ACC)[1], as well as joining a cooperative, plays an important role in agricultural enterprises, particularly regarding the equipment used. It has also been shown that DTPs can promote industrial agriculture and increase dependence on chemical inputs. Ecologization trajectories are linked to economic models, which leads us to our first proposition concerning the links between the use of DTPs and ecologization:

1 *Entreprise de travaux agricole* (ETA), in French.

– *Proposition 1*: the use of DTPs is associated with weak forms of ecologization and industrial-oriented economic models. Digital technologies for information and communication (DTCs) are changing the flow of information and knowledge exchange and opening up new possibilities for co-creation and sharing. Through its technical, social and political dimensions, ecologization requires changes in the creation and circulation of knowledge, particularly concerning agroecological models. DTCs can facilitate these knowledge transformations and promote strong ecologization.

– *Proposition 2*: the use of digital technologies is associated with stronger forms of ecologization, which rely on a redesign of production systems. However, digital technologies can, on the contrary, lead to the commodification, concentration and industrialization of information and knowledge. This could lead to a loss of skills and dependencies on trajectories that encourage the continuation of conventional agricultural systems. Moreover, strong ecologization requires learning that remains based on the exchange of local knowledge and farm experimentation.

5.1. Diversity of technologies and uses of digital technology

Before delving into the detailed analysis of digital technology use on farms, we present our sample here and describe the overall adoption of digital technologies, as well as a breakdown of usage for some key technologies (Schnebelin et al. 2022b).

5.1.1. *Sample description*

Our sample covers a diversity of farms (n = 98), as shown in Table 5.1. The utilized agricultural area of the farms varies from 9.5 to 570 ha, with an average of 162 ha, which is higher than the regional average of 59 ha. This difference can partly be explained by the inclusion, in the official census, of farms managed by part-time farmers or retirees. We chose not to include these categories in our sample, because they do not have a significant economic weight. The standard gross production (SGP) and labor were therefore also higher in our sample than the regional average, with great diversity. The respondents were between 24 and 67 years old.

Table 5.2 shows the distribution of the population defined by certain qualitative variables. Mixed farms are those where only a portion of the crops is grown organically, the other crops not being certified organic.

	Sample mean	Minimum	Maximum	Regional average (Agreste 2016)[2]
Utilized agricultural area (UAA), ha	162.1	9.5	570	60
Standard gross production (SGP), €	243,157	5,559	1,053,895	€58,070 (COP 2016)
Gross operating surplus (GOS), € (58 responses)	55,452	9,572	180,000	–
Total workforce	1.9	1	6	1.4 (2016, all OTEX) 0.88 (2016, COP)
Age	44.1	24	67	53 (2016, all OTEX)

Table 5.1. *Descriptive variables for the sample*

		Number	%
Crop certification	**Organic**	28	28.6
	Conventional	58	59.2
	Mixed	12	12.2
Education	**Bac 2**	31	31.6
	Bac + 2	42	42.9
	Bac 2	25	25.5
Gender	**H**	89	90.8
	F	9	9.2

Table 5.2. *Qualitative descriptive variables for the sample*

5.1.2. *Adoption of digital technologies in our sample*

Tables 5.3–5.6 summarize the use of digital technologies by the farmers in our sample. Table 5.3 lists the main digital production technologies (DTPs). It highlights that GPS guidance technologies and farm management software are widely used. Sections 5.1.3 and 5.1.4 describe these technologies and their uses in detail. Other technologies, such as section control, connected weather stations or fertilizer modulation technologies (description in section 5.1.5), are regularly used on the farms surveyed. Tables 5.3–5.5 provide some figures on internet usage to obtain technical information. They show that farmers make extensive use of the internet to obtain technical information, particularly specialized websites and social networks. Far from being a simple consultation of information, the internet has allowed 45% of

2 See: https://agreste.agriculture.gouv.fr.

the farmers surveyed to make changes in their practices. Section 5.1.6 returns to the way farmers use the internet.

Uses a GPS	60%
Uses autoguidance	44%
Uses section control technologies	39%
Has a connected weather station	32%
Uses modulation cards	33%
Applies variable-rate nitrogen	27%
Applies variable-rate seeding	6%
Equipped with connected tensiometers	14%
Yield controller	18%
Uses decision support tools for treatments	11%
Uses connected irrigation technologies	15%
Uses management software	63%

Table 5.3. *Use of technologies*

Never	3.1%
Rarely	16.5%
From time to time	15.5%
Often	41.2%
Very often	23.7%

Table 5.4. *Frequency of internet use for collecting technical information*

Uses social networks	53%
Uses Facebook	24%
Uses YouTube	24%
Uses Twitter	5%
Uses forums	10%
Uses specialized websites	81%
Uses technical institute websites	30%
Consults the agricultural press online	17%
Uses organic farming websites	12%
Uses conservation agriculture websites	12%

Table 5.5. *Technical information on the internet*

Participates in technical discussions on the networks	19%
Changed their practices thanks to the internet	45%
Including change of practices on the cover crops	51%
Including organic changes in practices	12%

Table 5.6. *Type of use*

5.1.3. *Use of farm parcel management software*

The farm parcel management software is a software designed to manage one or more agricultural operations (see Figures 5.1 and 5.2). Based on a map of the farm, it is designed to be able to track its interventions on each plot (tillage, input application, harvesting, etc.). The software can also be used by an agricultural advisor who will record their advice directly in the software. It can also serve as an interface to other digital services.

In our sample, 65% of farmers use farm parcel management software.

A technology adopted on average in 2013, made known through the cooperative, the suppliers, the chamber of agriculture, neighbors, accountants or directly via the company Isagri. The main provider is Isagri: 41%. Next come the cooperatives and suppliers (28%) and the chambers of agriculture. The median annual subscription price is €370. Prices may vary depending on the options subscribed.

Functions: the first cited function is regulatory traceability for plant protection treatments. The software is used by 34% of farmers to make their fertilization plan (regulations on fertilization). A quarter of farmers use the software for the economic management of their farm: calculating margins, costs, etc. Several report not doing this economic management with their software because it is too long and too complicated. Furthermore, the software is also used for technical management (34%): organizing crop rotations, keeping a history of technical itineraries, etc. About 19% of farmers use it with their technician to record their recommendations in particular; 13% mention use for the CAP. Farmers mention other less frequent functions: stock management, management of contracts or specifications (HVE, GlobalGAP, etc.), client management within the framework of their ACC, management of working time and supervision of employees, or even organic certification.

Link with practices: the majority of farmers (87%) report that using the management software has not led to any changes in their agronomic practices. Note that 12% have changed their practices because of the software, whether through a reduction in applied doses, a change of product or a change in technical itinerary.

These changes occur following analyses of their own history, either alone, alongside other farmers or with technicians.

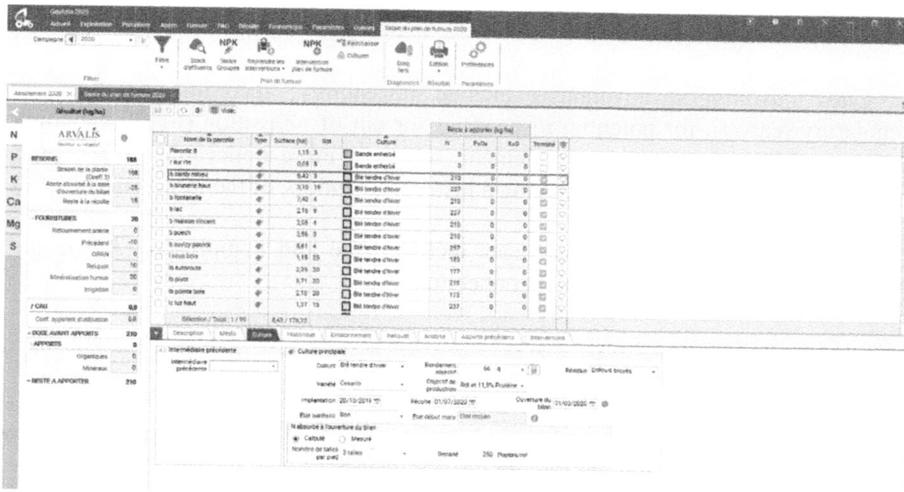

Figure 5.1. *Land parcel management software, visualization of technical interventions. For a color version of this figure, see www.iste.co.uk/schnebelin/ecologization.zip*

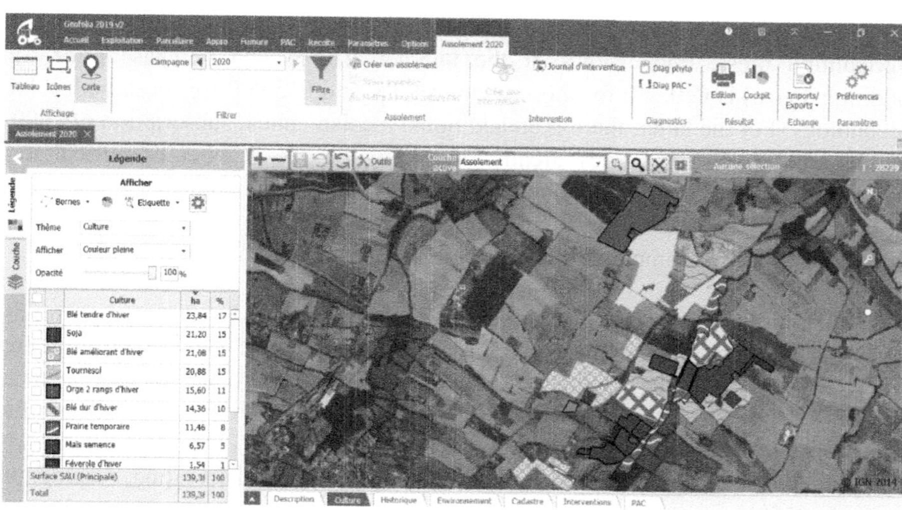

Figure 5.2. *Land parcel management software, parcel visualization. For a color version of this figure, see www.iste.co.uk/schnebelin/ecologization.zip*

Link with the organization: similarly, 87% of farmers report no change in their organization. In fact, many continue to do their record-keeping on paper before entering it into the software. Only 11% report changes, moving toward better economic management of their farm.

Cited advantages: advantages include traceability; recording practices for regulatory reasons, for potential inspections, or out of obligation (61%); economic analyses (25%); agronomic management, which is more noticeable (37%); time-saving and comfort (16%); and having a history, a memory of the farm's data (33%).

Cited disadvantages: disadvantages include time and complexity (39%), cost (13%), and dependence on internet connection.

5.1.4. *Use of guidance*

Guidance refers to the use of a geolocation system on agricultural machines to guide or automate driving. There are several categories of accuracy, from 20 cm down to 2 cm. The 2 cm accuracy is based on the RTK system, which requires a paid subscription.

In our sample, 61% of farmers use a guidance technology, 44% use automatic guidance and 25.5% have RTK technology.

The technology was adopted on average in 2013, known of via peers, dealers, the agricultural press, trade shows or the internet. The median purchase price (or additional cost) is €10,750 for those without RTK and €15,000 for those with RTK technology. For the latter, there is an additional median annual subscription of €600. The initial investment varies greatly depending on the number of agricultural machines equipped. The suppliers are mainly agro-equipment companies. There are also players dedicated to digital agriculture (Isagri, Trimble) or specializing in GPS (Innov GPS). In our sample, one farmer built this technology himself.

Functions: a large number of farmers (39%) use guidance for all crop operations requiring the tractor. Thus, farmers either have multiple GPS devices on different tractors or prioritize the tractor that is equipped, while the other tractors become secondary or are used for other tasks. Other farmers mention one or more specific tasks for which they use guidance: sowing and planting (29%), fertilization (27%), spraying (27%), tillage (15%), weeding or hoeing (15%) and harvesting, reaping, or mowing (8.5%).

Links with existing practices: most farmers report not having changed their practice with the use of guidance. The main changes are linked to conservation agriculture techniques, direct seeding under cover or the use of strip-till. Or changes with regard to hoeing and fertilization. Other changes may be linked to the use of other technologies combined with guidance (e.g. variable-rate application).

Link with organization: most farmers state that the use of guidance has not changed their organization. However, some report working longer hours and working at night (8.5%). Working more land or changing the organization of labor is also mentioned. Some also highlight changes in the use of agricultural equipment, for example, using only one tractor.

Figure 5.3. *Yield map created by combining yield monitor and geolocation. For a color version of this figure, see www.iste.co.uk/schnebelin/ecologization.zip*

Advantages cited: advantages include comfort, reduction of fatigue and arduousness (cited by 59% of farmers); precision, optimization and avoiding overlaps (45%); time savings and increased work rate (68%); saving money (on fuel, inputs, etc.) (41%), with fuel being the most frequently mentioned issue (25%); having professional, precise, regular work (12%); monitoring and adjusting the tool (16%); less soil compaction, reducing field passes (11%); and less equipment wear, collecting data and ensuring traceability, working at night.

Disadvantages cited: disadvantages include cost (32%); software errors and signal loss (37%); difficulties in using it, the need to train oneself and employees (12%); dependence (12%); the time spent to use this technology (8.5%); poor accuracy (3%); obsolescence; and data appropriation.

5.1.5. *Modulation use*

Modulation maps are maps created to characterize within-field heterogeneity, based on soil or vegetation analyses, and to recommend fertilization or seeding rates (see Figure 5.3). These maps can be used with a spreader or seeder adapted for modulation, which adjusts the input rate continuously according to the map's recommendations.

In our sample, 34% of farmers use modulation maps and 27.5% perform automatic fertilizer modulation with them.

The technology was adopted on average in 2016. Costs vary depending on the type of map: €9/ha/year on average for basic fertilizer modulation, and €100–€150/ha for maps created with soil analyses (valid for several years). The adapted equipment (console and dispenser) involves additional equipment costs (from a few thousand euros to several tens of thousands of euros). Maps are provided mainly by cooperatives and trading companies (75%), sometimes by seed companies (12%), or more rarely by a chamber of agriculture or a CETA (a technical exchange group).

Functions: fertilizer modulation is used mainly for wheat, barley, rapeseed and maize. Maps are obtained via satellite images, or sometimes through soil analyses, and a minority (3) create their own maps from their parcel management software maps and their knowledge of the fields. Some perform modulation manually, lacking the equipment for automatic modulation. Some adjust automatic modulation based on their knowledge, observations or objectives and can correct errors (e.g. taking into account weed biomass). Some modulate all applications, while others modulate only the last application. For initial applications, farmers increase doses on low-potential areas to homogenize the field. For the last application, farmers increase doses on high-potential areas to maximize yield and protein content.

Links with practices: most farmers report not having changed their practices, except for fertilization management (increasing or decreasing doses, modifying the number of field passes and application dates). Some farmers mention having increased the area of the crop concerned by the technology in order to make the investment cost-effective.

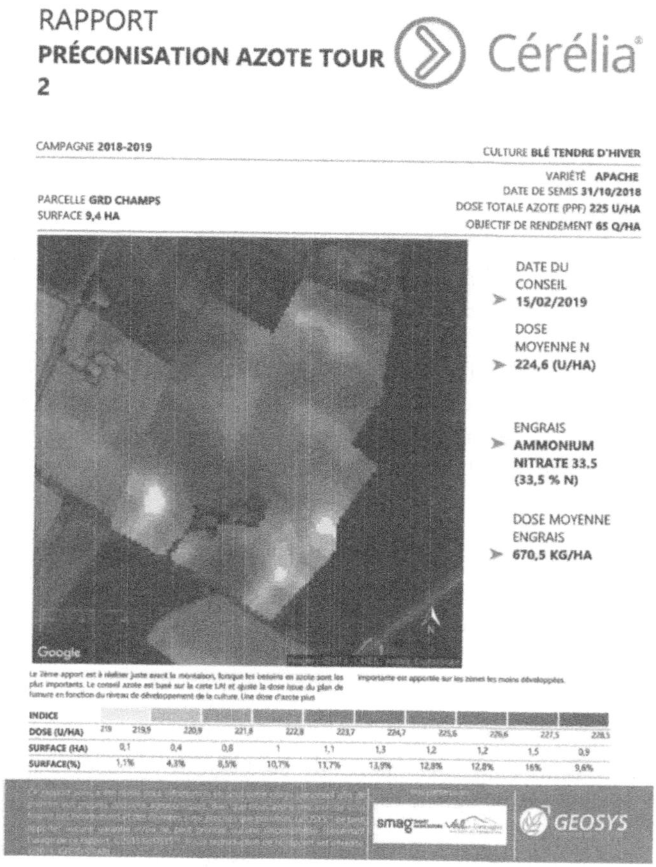

Figure 5.4. *Modulation map for nitrogen fertilizer application. For a color version of this figure, see www.iste.co.uk/schnebelin/ecologization.zip*

Link with the organization: the impacts on the organization mentioned are the management of fertilizer stocks – some farmers adjust fertilizer orders based on the recommended doses – and anticipation, in order to order the maps on time. This can cause problems during cloudy weather, which prevents the maps from being created.

Advantages cited: advantages include fertilizer savings (32%), although few are able to estimate the actual savings; yield optimization (21%); production quality (24%); regulatory justification of the applied dose (18%); ease of use (18%); and order management.

Disadvantages cited: disadvantages include cost (19%); time required (19%); lack of effect, gain, or savings (19%); the need to anticipate (16%); the availability of maps when needed (13%); errors (13%); and problems with suitability to the terrain.

5.1.6. *Internet usage*

In our sample, 65% of farmers report using the internet as a source of technical information often or very often, 15% occasionally and 20% never or rarely.

To find this technical information, 81% visit specialized websites, notably the websites of technical institutes (30% of surveyed farmers), websites specialized in organic farming (12%), or in conservation agriculture (12%). Note that 53% go to social networks, mainly Facebook (24%) and YouTube (24%), or specialized forums (10%).

Note that 45% report having already made changes in their practices thanks to the internet. Half of these changes concerned cover crops, and 12% concerned changes related to organic farming.

5.1.6.1. *Advantages and disadvantages*

The internet allows farmers to access information that they do not necessarily have access to within their local network, notably information on cover crops, conservation agriculture or organic farming.

Note that 35% of farmers report disadvantages related to technical information on the internet. Among the problems mentioned are mainly the reliability of information and the fact that the information is not adapted to their context. Farmers also point out that it is difficult to navigate due to the abundance of information. In addition, several mention the issue of data protection.

Furthermore, in-person exchanges with their colleagues, their technicians or training sessions are mentioned as indispensable and complementary.

5.2. Profiles of use

To go beyond the use of technologies one by one, we carried out a classification to construct digital usage profiles for the two types of technologies identified (Schnebelin 2022b). The classification method resulted in three groups for production technologies (DTPs) and three groups for communication technologies (DTCs), as shown in Figure 5.5.

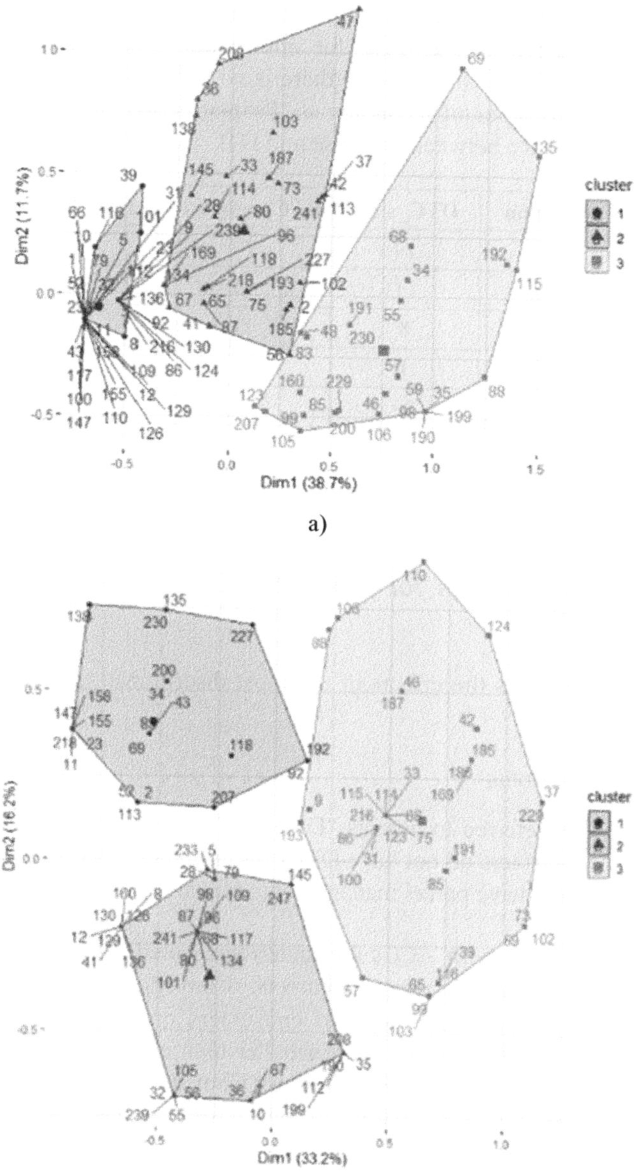

Figure 5.5. *Classification of farmers according to their use of digital technologies for production (a) and for information and communication (b). For a color version of this figure, see www.iste.co.uk/schnebelin/ecologization.zip*

Two profiles were constructed to characterize the two types of technologies: those for production (DTP) and those for information (DTC). By crossing these profiles (see Table 5.7), we showed that there is no significant correlation between the uses of these two types of technologies. Pearson's test enabled us to reject a hypothesis of dependence between DTP use and DTC use in farms.

	DTC_non	DTC_sites	DTC_networks	DTC N/A	Total
DTP_non	8	19	10	0	37
Average DTP	6	11	13	1	31
DTP_intense	8	8	11	1	28
DTP N/A	0	1	1	0	2
Total	22	39	35	2	98

Table 5.7. *Number of farmers in each of the digital usage profiles*

5.2.1. *Digital technologies for production associated with productive models*

Table 5.8 summarizes the effects of the most discriminating variables between the three groups.

5.2.1.1. *DTP_non profile*

The first profile, referred to here as "DTP_non", represents 39% of the sample. In this group, most farmers do not have DTPs for production. However, slightly less than one-third of them have parcel management software.

In this profile, farms with ACCs are underrepresented, as are those cultivating seeds or having commercialization contracts. Conversely, organic farms, sole proprietorships, farms selling directly to consumers or farms with livestock are overrepresented. Farms in this group are smaller than average, with lower acreage and gross margins, fewer annual loans and less labor. However, they do not have a lower gross operating surplus (GOS). Input costs per hectare and yields are significantly lower.

5.2.1.2. *DTP_average profile*

The second profile, named "DTP_average", represents 32% of the individuals in the sample. It includes those who use some DTPs, but do not have all precision agriculture technologies.

Thus, almost all use guidance technologies (97%) or autoguidance (63%), and a large majority (87%) use parcel management software, but none use variable rate application. Furthermore, between 20 and 40% use section control technologies, decision support tools for their treatments, weather stations, connected tensiometers, or connected irrigation technologies.

The farms in this profile generally resemble the sample average, but are distinguished by higher irrigation, more salaried labor, higher fertilizer costs per hectare and less direct sales. The organic farms in this group (N = 8) have higher GOS and operating profit, loan repayments, land area and irrigated shares, and sell more through cooperatives.

5.2.1.3. *DTP_intense profile*

The third profile, DTP_intense, represents 29% of the sample and is characterized by intensive use of DTPs. Many use precision agriculture tools: guidance systems (96%), variable-rate fertilizer application (92%), connected weather stations (65%) and yield monitors (46%). Note that 23% use variable-rate seed application.

The farms belonging to this group are significantly larger than those in the other two profiles, both in terms of area and economic size. They also have a larger workforce (total and salaried). These farms have higher input costs, but also higher yields, resulting in a higher GOS per hectare.

This profile overrepresents farms with agricultural contracting services (ACCs), mixed farms, those growing seeds and those with a sustainable farming strategy. It includes fewer certified organic farms and fewer of those with intercropping systems.

Variables	DTP_non	DTP_average	DTP_intense
Structural variables			
Utilized agricultural area (UAA)	– ***	–	+ ***
Owns an agricultural contracting company (ACC)	– ***	–	+ ***
Standard gross production	– ***	–	+ ***
Total labor requirement	– *	–	+ **
Salaried labor requirement	– **	+ ***	+ **
Annuities	– ***	+ *	–
Gross operating surplus (GOS)	–	–	+ **
Has livestock	+ ***	–	– *

Individual variables			
Digital skills	3: + **	3: – *	–
Socioeconomic environment	–	–	–
Serves as a cooperative advisor	– **	–	+ *
Marketing through trade	– **	+ **	–
Marketing seeds	– ***	–	+ **
Having marketing contracts	– ***	–	–
Member of a knowledge sharing group	– **	–	–
Member of an organic farming group	+ **	–	– *
Member of a machinery sharing group (CUMA)	–	+ *	–
Goes to agricultural fairs	– ***	–	+ ***
Makes direct sales	+ ***	– **	–
Agronomic practices			
Type of input management (1 conventional, 2 reasoned, 3 organic)	1: – ** 2: – * 3: + ***	–	1 2: + ** 3: – ***
Biodiversity cultivated (1 grassland, 2 cereals, 3 diversity, 4 legumes)	1: + *** 2: – *	1: – **	–
Soil management type (1: tillage , 2: deep tillage, 3: shallow tillage, 4: no-tillage)	2: – * 3: + **	–	–
Model (conventional, organic or mixed)	Organic + *** Conv – **	Organic Mixed – * Conv + *	Organic – *** Mixed + *** Conv
Practices intercropping	+ **	–	– ***
Fertilizer input per ha	– ***	+ ***	+ ***
Pesticide application per ha	– ***	+ *	+ ***
Soft wheat yield	– ***	–	+ ***
Durum wheat yield	– ***	+ ***	+ *
Share of irrigated land	– ***	+ ***	–

Table 5.8. *Summary of interactions between farm variables and DTP usage profile (+: positive effect on the probability of belonging to the profile; –: negative effect; *p < 0.1; **p < 0.05; ***p < 0.01)*

5.2.2. *Digital technologies for information and communication associated with individual practices*

Table 5.9 summarizes the effects of the variables that most strongly distinguish the three groups.

5.2.2.1. *DTC_non profile*

The first profile, DTC_non, is characterized by less frequent internet use for information. Two-thirds never, rarely or only occasionally use the internet to search for technical information. Three-quarters do not use social networks and do not consult specialized websites. Almost all do not use YouTube or the websites of technical institutes. However, 45% consult the agricultural press.

This profile is characterized by fewer individuals with higher education (beyond 2 years of post-secondary studies) or belonging to a CETA, but more individuals receiving advice from cooperatives or having lower computer literacy. The farms have higher than average pesticide costs per hectare, less no-till farming and more plowing.

5.2.2.2. *DTC_sites profile*

The second profile, DTC_sites, is defined by the use of specialized websites to search for information. They do not use Facebook, YouTube or online agricultural press, but all go to specialized websites, particularly the websites of technical institutes (36%).

This profile is characterized by fewer farmers cultivating seeds, but more using no-till or simplified cultivation techniques. On average, they have higher proportions of legumes, lower pesticide use and higher yields of irrigated maize and durum wheat. These individuals are older than average, they are more numerous in the group of those who have an intermediate level of digital skills, and fewer belong to a union.

5.2.2.3. *DTC_networks profile*

The third profile, DTC_networks, is characterized by the use of social networks for professional purposes (100%), particularly Facebook and YouTube. They use the internet often to very often for 91% of them. They also consult specialized websites (89%).

This profile is distinguished by more EARLs (limited liability farms) and more farms with irrigation. They have been established more recently than average and are younger. More of them are in the group of those with the highest digital skills and those expressing a negative opinion about the economic health of their farm.

Variables	DTC_non	DTC_sites	DTC_networks
Individual variable			
Age	–	+ **	– ***
Installation date	– *	–	+ ***
Digital skills	3: + **	1: – ** 2: + ***	1: + ** 2: – *
Training	> Bac + 2 – ***	> Bac + 2 *	–
Perception of the economic performance of their farm	–	–	Negative + **
Have children	–	+ **	–
Socioeconomic environment			
Advise in a cooperative	+ *	–	–
Market seeds	+ *	– ***	–
Contracts	–	– *	–
CETA membership	– **	+ *	–
Membership to a union	–	– **	+ *
Member of a knowledge-sharing group on simplified cultivation techniques or no-till farming	–	+ *	–

Agronomic practives			
Soil management type (1: tillage, 2: deep tillage, 3: shallow tillage, 4: no-till)	1: + ** 4: – **	–	–
Is in no-till farming	– **	+ **	–
Legume share	–	+ **	–
Pesticide application/ha	+ **	– **	–
Durum wheat yield	– *	+ *	–
Irrigated portion	–	– **	+ *

Table 5.9. *Summary of interactions between farm variables and DTC usage profile (+: positive effect on the probability of being in the profile; –: negative effect; *p < 0.1; **p < 0.05; ***p < 0.01)*

Compared to the clusters concerning production technologies, fewer variables are significantly different from one group to another. This confirms the fact that we have less pronounced profiles.

5.3. Digital uses and ecologization

5.3.1. *DTP, industrialization and weak or symbolic ecologization*

5.3.1.1. *DTP and enlargement*

A large economic dimension emerges as one of the main factors explaining the use of DTPs. First, it facilitates access to these technologies and investment in them. The economic dimension sufficient for investment can be achieved through farm expansion, through collective investment (relatively little used in our sample), or through an "expansion via service provision". Nguyen et al. (2020) show that subcontracting is a way to access the use of this type of technology without investing. We complete this analysis by showing that subcontracting can be seen as a way to invest. It is one of the ways to engage in a logic of expanding cultivated areas, income and thus investment capacity.

Second, these DTPs also respond to farm-specific needs and offer additional advantages in using these tools. They sometimes become indispensable for managing work: organizing labor, knowing what has been done on which plot, standardizing practices and managing logistics and traceability. For example, digital tools are used to record and track work performed on different plots, or to communicate technical itineraries to tractor operators. The use of GPS guidance also allows increasing working speed, but also working time by reducing fatigue related to tractor driving and by enabling driving at night, as described by Baptiste:

> I don't know if I actually saved time. Because as soon as you save time, you have the time to do something else and do more. It's a cycle. [...] My great-grandfather had 16 ha. My grandfather had 200. My father 400–450. Now we do 800 or 900. (Baptiste, intensive DTP user)

The enlargement of the economic dimension can then be seen not only as a cause but also as a consequence of DTP use, which allows overcoming obstacles to expansion and sometimes increasing labor productivity. Digital technologies also respond to the traceability and standardization needs of ACCs, which sometimes operate on dozens of different farms.

The economic dimension influences the adoption of these technologies, which encourage the continuation of the expansion trajectory of these models. This reinforces the hypothesis put forward in many social science studies, showing that digital technologies, and more specifically those of precision agriculture, tend to favor and lock in the dominant industrial agricultural system (Carbonell 2016; Bronson 2019; Rotz et al. 2019b; Clapp and Ruder 2020).

5.3.1.2. *Role of value chain organizations in the use of DTPs*

This study highlights the crucial role of value chain actors, such as agricultural cooperatives and advisory services, in the dissemination and adoption of DTPs. We confirm the role of cooperative advice or membership in farmer groups in the adoption of DTPs (Barnes et al. 2019).

We primarily demonstrate that seed crop production, which is associated with specific marketing and advisory services, promotes or even imposes the use of DTPs, and as do marketing contracts. Downstream companies (cooperatives, traders and seed producers) encourage or impose the use of these technologies, via contracts, to standardize and better control their supply.

5.3.1.3. *DTP to optimize or justify conventional agricultural models*

These technologies are based on input optimization. This limits their relevance for farmers who have strategies other than this optimization, for example, those in organic farming, as illustrated by the testimony below:

> The other farmers who have GPS [...] they save diesel, seeds, fertilizers, herbicides. Because they apply just what is needed. So for them, maybe it results in savings that allow them to pay for the system. But for me, I will only save on diesel, and also save time. (Thomas, DTP_non)

Moreover, a large proportion of organic farmers say that DTPs are not suitable or even contradict their production system. Some farmers transitioning to organic have stopped using parcel management software or decision-support tools that were no longer useful to them.

In the DTP_non group, the SGP is lower, but the GOS is not. This may mean that some ecologization trajectories seek other segments of value-added optimization than those proposed by precision agriculture. This could be through processing and/or direct sales, through the optimization of ecosystem services, etc. DTPs are consistent with farms that have a strategy based on increasing productivity and efficiency and on maximizing yield (Bronson 2019), but less so with those having strategies based on reducing external inputs through agronomic techniques or on diversification rather than specialization.

However, the effects of technologies on optimization are questioned by some users, particularly regarding variable rate technology. Some report that variable rate technology does not lead to savings, and some even stop using it and revert back. Some question the effects on yields and input consumption. This relates to criticisms of DTPs, which appear more as tools for "symbolic ecologization", that is, to justify practices, rather than tools for real change and optimization of practices (Wolf and Wood 1997).

5.3.1.4. *Uses of DTP associated with important changes in ecologization trajectories*

The results show that choices in terms of agronomic practices and choices in terms of tools influence each other, with DTPs favoring optimization practices over redesign practices. However, this must be nuanced. Associations are also observed between more radical changes in practices and the use of DTPs: in the case of Louis, a change in practices led to his use of DTPs. Conversely, for Baptiste, the use of DTPs contributed to the implementation of new agronomic practices.

The implementation of sustainable soil management techniques led farmers toward the use of DTPs, especially precision auto-guidance, as Louis testifies:

> It is because we changed agronomic practices that we came to RTK. Since we work a lot with cover crops […] when we sow maize into cover crops that are a bit tall, it is good to have guidance. (Louis, DTP_moderate)

Moreover, mixed farmers are overrepresented in the DTP_intense profile. We observe farmers who have used DTPs to move toward organic farming, as illustrated by Baptiste below:

> If I hadn't had guidance in organic farming, I might not have switched, that's certain. Because we have power, we have width. So we have wide tools. Now we are in organic, we redo all the fields in 8.20 m with guidance, it's great. (Baptiste, TNP_intense)

It can be hypothesized that digital tools allow farmers who were conventional to move toward organic while keeping a somewhat similar working mode: organic fertilizers instead of chemical fertilizers, pesticides authorized in organic farming rather than conventional products and mechanical weeding with precision hoeing rather than chemical weeding.

Thus, DTPs would be consistent not only with efficiency strategies, but also with substitution strategies. The overrepresentation of mixed farmers must be put in perspective with the fact that these farmers operate on larger areas and often have ACCs. They fit into a more industrial model.

Digital technology appears as a factor favoring a stronger form of ecologization: the development of organic farming on large farms. This echoes a political and academic debate on the conventionalization of organic farming (Stassart and Jamar 2009; Darnhofer et al. 2010): Could digital technology be an accelerator of this conventionalization, to the detriment of more radical organic farming on small farms?

Moreover, some uses of DTPs are sometimes induced by other practices that can be associated with agroecology. For example, some farmers use GPS guidance to implement under-cover sowing techniques. This technique consists of sowing a crop while another plant, called a "cover crop", is already in place. It is promoted as a sustainable soil management technique, but also to limit the use of pesticides and labor time.[3] Guidance is used to compensate for the loss of visibility caused by the presence of the cover crop.

Others use digital tools to perform mechanical weeding, with precision guidance or with the help of cameras. These technologies allow farmers to weed as close as possible to the crops in place. One farmer also used guidance for implementing strip-cropping, a technique that consists of dividing a field into strips that alternate two or more crops, which limits soil erosion and reduces fertilizer requirements (Francis et al. 1986).

5.3.2. *DTC, knowledge and ecologization*

5.3.2.1. *Uses of DTC associated farmers' individual characteristics as well as with their integration into a socioeconomic environment*

The adoption of DTCs is strongly influenced by individual farmer characteristics, such as training level, age, skills and interest in this type of technology. However, economic variables are not significant in our sample, unlike the results of Michels et al. (2020), which focused specifically on smartphone adoption.

The socioeconomic environment is also linked to the use of DTCs through correlations or oppositions, depending on the type of organization concerned. Membership in a cooperative or advisory services via seed companies is negatively associated with the use of DTCs. This can be explained by the fact that technical itineraries for seed crops are not chosen by the farmer but imposed by the seed companies. Production contracts facilitate the exchange of knowledge (Cholez et al. 2020), so there is no need to seek additional technical information. Similarly, advice provided by cooperatives is considered sufficient by some users. On the other hand, membership in a CETA or in farmers' groups is positively correlated with the use of DTCs.

Membership in farmers' groups can indicate an interest in agronomic and technical innovation, and these groups also provide complementary sources of information, as noted in the interviews: seeing what is being done elsewhere versus seeing what works locally, obtaining a large amount of information versus having precise and field-adapted knowledge. Membership in a union is positively correlated with the use of social networks.

3 See: https://dicoagroecologie.fr/encyclopedie/semis-direct-sous-couvert-vegetal/.

5.3.2.2. *Internet: easier access to (limited) information on new practices*

According to the interviews, the internet appears as a source of information to deal with atypical situations or to gain knowledge not available in the local network. It is also a source of inspiration to try new practices and a means of agronomic monitoring. This goes from adjusting practices (to face new conditions related to climate change, for example), up to the search for information on more radical changes in practices, notably because the knowledge sought is not available in the usual network of information sources, as testified by Baptiste:

> The cooperative was a bit behind on that, so I went to see what was being done quite a lot. Also the mixtures. And so I would purchase, I would make my own little mixes. (Baptiste DTC_networks)

Sometimes the notion of exchange, participation and being oneself a source of knowledge is added to this. However, the problem of reliability or relevance of information in relation to local pedoclimatic conditions is often mentioned as below:

> Well, these are more like experiments, but then again what works for some will not necessarily work here. (Corentin)

Our second proposition is also confirmed, but with some limitations. Using the internet is both a cause and a consequence of changes in practices. The internet is not necessarily the preferred form of access to knowledge, but allows to fill a lack of agroecological information in the classic network (advice, neighbors, family) (Lucas 2021). Knowledge from the internet does not seem to be able to completely replace that acquired orally, notably for ecological, localized and adaptive knowledge (Burton and Riley 2018). Moreover, it is often farmers involved in local knowledge-sharing groups who mobilize technical information via the internet the most. Thus, DTCs allow to combine a diversity of information sources, rather than replacing other forms of knowledge exchange.

5.3.3. *Trajectories*

Qualitative perspectives complement the quantitative description of profiles of digital technology use. They make it possible to demonstrate trajectory mechanisms, in particular processes of nonlinearity.

Firstly, the process of digitalization at the level of individual farmers' decisions is not linear. Phenomena of technology experimentation followed by their abandonment can be observed. Second, farmers do not follow a single adoption curve. The categorization of "pioneers" or "laggards" is not relevant here. Not adopting a technology does not necessarily imply resistance to change, but may be a coherent choice corresponding to the farmer's practices and objectives.

The adoption and use of a technology are rather the result of a production model interacting with a socioeconomic system that encourages or imposes these technologies.

Finally, we observed reinforcement mechanisms: the use of DTPs further promotes trajectories of industrialization and expansion of farms, which in turn favors the use of DTPs. This leads to path dependence mechanisms that reinforce dominant agricultural production systems.

Our research also confirms the major role of intermediaries, such as advisory services, training organizations and value-chain actors, in the use of digital technologies and agricultural trajectories. Regulation, agricultural policies and private standards also play a role in digital uses and contribute to mechanisms of path dependence.

Overall, the results of surveys conducted among farmers show that the adoption and use of digital technologies do not appear as the result of an individual "pioneer" behavior, but depend above all on agricultural production models, that is, agricultural practices, labor and the economic structure of farms, in interaction with their socioeconomic networks of advice and marketing, which propose, encourage or even impose these technologies. These results suggest trends and dynamics in the trajectory of the digitalization of agriculture.

5.4. Digitalization: an imposed, desired or contested trajectory?

This cross-sectional analysis highlights factors that influence the trajectory of the digitalization of agriculture in France. It appears to be both imposed on farmers, who are forced to digitalize, desired by professionals seeking a better future, and contested on technical, economic, political and environmental grounds.

5.4.1. *The influence of regulation, institutions and public policies*

The development of digital technology across the agricultural sector as a whole is supported by public innovation policies. At the individual level, surveys of farmers have also highlighted the role that institutions play in digitalization. First, regulatory requirements have served as a starting point for the use of digital tools on many farms. Accounting obligations led to new management procedures, followed by the digitization of public services and reporting processes (e.g. common agricultural policy online declarations became mandatory in 2016), as well as by various regulations, such as traceability requirements for practices like fertilizer or pesticide applications. Our results underline the importance of environmental regulations, which have driven digitalization, as previously noted in the literature (Koutsos and Menexes 2017; Oui 2021b). Requirements for fertilization plans and the traceability of input use appear as key triggers for adopting a central digital technology: farm management software.

The increasing complexity of pesticide regulations adds to these requirements: digital tools have become a way to access this information that is constantly changing. Farm management software allows farmers to understand regulations and implement them. This also leads to certain workaround strategies: when alerts indicate that entered information does not comply with regulations, some farmers may adjust the data they input into the software. Variable-rate technology also serves regulatory purposes, as it allows farmers to justify plant "needs" and thus fertilizer applications that may exceed the standard authorized threshold. Connected weather stations also enable regulatory traceability. For instance, applying sanitary products is prohibited when wind speeds are too high. Recording data from local anemometers allows farmers to justify applying a product at a time when official weather data suggested the wind was too strong but local measurements indicated acceptable conditions. In this sense, the development of digital technology can be seen as a response to regulatory requirements. However, environmental regulations appear to be a demand for traceability and justification rather than for actual changes in farming practices.

Moreover, the regulation of agricultural practices is not solely determined by public norms and policies. Contemporary regulation of the agri-food sector results from the interplay of both public and private regulatory mechanisms (Touzard and Labarthe 2018). Technical standards imposed by the agri-food industry or buyers can thus strongly influence production practices and their justifications (Cholez et al. 2017). Following the same logic of quantification, standardization and traceability encouraged by public regulations, our results show that private standards can also promote – or even require – the use of digital tools. For example, contracts in the wheat sector with agri-food companies mandate the use of variable-rate fertilizer technology. Similarly, seed companies contractually require seed producers to use variable-rate fertilizer or irrigation technologies. This confirms hypotheses in the literature that private standards can become drivers of digitalization (Prause et al. 2020).

Several factors contribute to the development of this standardization and the associated digitalization. Digital tools provide valuable data for both upstream and downstream industries. These data can facilitate tighter production control, the sale of data-based services to farmers as well as production forecasting and management. For downstream industries, data can be monetized – for example, through labels or QR codes – to segment markets, increase product value, create a differentiation advantage and build credibility through traceability (Valceschini 2006). Traceability here is not necessarily linked to changes in practices but becomes a marketable quality in itself, which could be described as "traceability quality" or "informational quality".

Thus, digitalization partly results from changes in both public and private regulatory frameworks. Consequently, in interviews, several farmers perceived digitalization as an obligation rather than a technological choice. While all innovation system organizations identify regulatory evolution as a driver of digitalization, this appears even more pronounced at the farm level for conventional farms. Conventional production involves the use of products that demand stricter traceability. Indeed, 22% of conventional farmers see digital technologies as advantageous for meeting regulatory or traceability requirements, compared to only 7% of organic farmers. Moreover, downstream value chains encouraging digital adoption mainly concern conventional farming.

5.4.2. *Farmers committed to and driving the development of digital technology*

However, digitalization on farms is not driven only by regulations. It is also implemented and developed for other objectives and carries hope for some farmers. First of all, the majority already see advantages in it: time savings (mentioned by 33% of farmers in the sample), access to information or communication (26%, but 40% among those in organic farming (organic or mixed farming) and simplicity (23%, 11% in organic farming). Other advantages mentioned include regulatory compliance (19%, 7% in organic farming), comfort and convenience (19%), precision and technical performance (16%), cost savings (10%, 4% in organic farming), sales, production (yield or quality) and even the analysis and understanding of their system. These advantages match those highlighted in studies in Germany (Knierim et al. 2019) and more broadly in Europe (Kernecker et al. 2020). Moreover, some surveyed farmers see potential in the development of digital tools in their sector. For some, digital technology "is the future and will offer endless possibilities for agriculture"[4]. They believe it can improve farm profitability, working conditions and environmental protection. Conversely, some (16%) see no

4 Mentioned by a farmer in DTP_average and DTC_networks profiles.

potential, feeling that "progress in this field has already gone too far"[5]. For many, digital tools could bring benefits to agriculture, but with limits. They could help optimize and save inputs, providing both economic and environmental benefits, and improve working conditions by increasing comfort or reducing repetitive tasks, provide access to information, enhance agriculture's image in society, better analyze farm systems and enable agronomic progress. Some even mention robotics as a potential benefit for agriculture. These potential benefits are sometimes weighed against longer term consequences that they can bring on: for example, time saved through digital tools could be used for other activities, and improved work comfort linked to digital tools could allow longer working hours.

Digital technology therefore attracts some farmers, some of whom are interested in the technological aspect of agriculture, the idea of "being modern", or because they "like gadgets". Some actively invest themselves in this development. For example, several respondents volunteered to test out technologies for their cooperative, confirming similar dynamics as those observed in Germany (Knierim et al. 2019). Another type of investment in digitalization is farmers' role in communication and training with colleagues on behalf of digital agricultural companies. In exchange for free access to software, these farmers promoted it and helped others implement it.

Perceived advantages of digital technology differ partly between organic and conventional farmers, especially regarding cost savings, simplicity, traceability and precision (more often mentioned by conventional farmers) or access to information and sales (more often mentioned by organic farmers). However, perceived potential is relatively similar.

Despite the appeal of these technologies, some acknowledge that they can sometimes be superfluous. Others, disappointed by tests conducted by the cooperative, even decided to abandon certain technologies. Criticism, and sometimes opposition, to digitalization is expressed by some farmers.

5.4.3. *Opposition to digital technology*

The majority of farmers surveyed see disadvantages in using digital technologies on their farms. The most frequently mentioned disadvantage is the cost of digital equipment (41%), which confirms findings in the literature on the topic (Knierim et al. 2019; Kernecker et al. 2020). Farmers also mention other inconveniences, such as the time required to use digital tools (21%), problems with bugs, breakdowns or compatibility (19%), and difficulties in repairing these tools (11%). They also see

5 Mentioned by a farmer in DTP_average and DTC_non profiles.

challenges in training and becoming competent (17%), as well as concerns over the appropriation of personal data or surveillance (10%). This development is seen as happening too quickly, and tools become obsolete rapidly. Additionally, some note that it "harms relationships and dialogue". Digital technology is sometimes seen as unsuitable for farm sizes, and, according to organic farmers, for local conditions or organic farming practices.

Farmers, whether they use digital technologies or not, perceive many risks associated with their development. The primary risk is related to data and is mentioned by one-third of respondents: data could be hacked or appropriated by digital companies, which can influence the decision not to use certain digital technologies (Briggeman and Whitacre 2010). As a result, several farmers (17%) cite risks of surveillance and control. Moreover, many criticize the dependency risks brought on by digital tools: dependency on digital technologies, as well as on networks, the agricultural industry or their cooperative. These dependency risks are mentioned far more frequently by organic farmers (about one-third) than by conventional farmers (about 10%). Farmers also highlight risks related to losing the ability to repair equipment, which becomes too complex with digital technologies, and that digital tools may become quickly obsolete or damaged by dust, humidity, etc. Some also perceive risks associated with the loss of know-how, fields-specific knowledge and observational skills, particularly the ability to avoid applying advice without personal reflection. These risks regarding control, repairing skills and knowledge loss are also documented in literature from England (Barrett and Rose 2020) and Australia (Jakku et al. 2019). In terms of sector development, some fear that digitalization may accelerate farm enlargement and concentration, harm agricultural employment and exclude parts of the farming population (those without the skills, or in less productive areas who cannot afford such technologies). Indeed, the cost of technology, as well as the debt and overinvestment associated with it, is seen as a major risk. Some, particularly organic farmers, highlight risks of relational loss, notably with advisors. Some organic farmers also mention environmental risks, related to the energy required to operate digital tools and the metals used in their construction. Farmers' statements also reflect risks of loss of purpose and changes in the nature of their work: digitalization could turn them into mere "executors" or "make them work even more time in the office". Regarding social media, risks such as agribashing and data dissemination are also frequently mentioned.

The perceived risks are thus quite diverse and are found among both conventional and organic farmers. However, organic farmers emphasize three risks more strongly: dependency, relational loss and environmental impacts due to energy use and materials. Beyond the individual opposition observed in our surveys, criticisms of agricultural digitalization are also expressed collectively by organizations and farmers' unions such as Atelier paysan (2021) or Confédération Paysanne (2021), as well as by environmental associations (Réseau Action Climat

2020), journalists (Tardieu 2017; Reinert 2022) and other actors in the sector (Aspexit 2021).

5.5. Contributions and perspectives

The field surveys conducted provide a view of the digitalization of agriculture “in the making”. This analysis of digital technology usage on farms reveals mechanisms that influence the transformation dynamics of the agricultural innovation system and provides a set of complementary elements to the analysis carried out in Chapter 4. In line with the results of our system-level analyses, the farm-level surveys show that the type of ecological approach affects the way digital tools are used. Indeed, certain digital technologies appear more advantageous for specialized, conventional farms. Thus, divergences can be observed in the use of digital technologies by farmers in organic versus conventional farming, which may be linked to differences in reasoning and logic (redesign and diversification versus optimization and simplification), the inputs used (organic or synthetic), regulations and the socioeconomic system. However, there are also expectations and perceived risks shared among all farmers. Farmers expect digital tools to provide time-saving and convenience, cost reductions and access to information and knowledge. At the same time, they are concerned about data appropriation by companies, surveillance and control, costs and the loss of the ability to repair equipment. Organic farmers additionally cite risks of dependency, relational loss and environmental impact. The analysis presented in this chapter shows that digital technology usage is also linked to other differentiating factors and seems to reinforce ongoing transformations such as farm enlargement, the increasing role of hired labor and subcontracting, and specialization. These uses thus strengthen divides between agricultural models and increase differentiation, potentially accentuating lock-in phenomena (Bronson 2019; Clapp and Ruder 2020).

However, the analysis of digital technology usage also reveals forms of de-compartmentalization or hybridization between paradigms. The use of DTCs, in particular, appears to form knowledge-sharing communities that transcend traditional organizational boundaries. Mixed profiles, combining organic and conventional farming and using many DTPs, also constitute a new form of hybridization. Technologies thus appear as an artifact that facilitates the hybridization of practices drawn from organic and conventional farming. Forms of bricolage and adjustment also create alignments between digitalization and strong ecologization on farms. The development of these interactions may contribute to gradual institutional change and could lead to new forms of hybridization between models. Conversely, this could also reinforce the integration of elements from alternative models into the dominant model in order to respond to critiques and strengthen its legitimacy (Beus and Dunlap 1990).

Finally, the results of this fieldwork once again highlight the role of intermediary actors. The analysis of farmers' ego-centered networks indicates their position within the innovation system and identifies organizations or actors influencing digital technology usage: marketing network companies such as agricultural cooperatives, advisory bodies such as chambers of agriculture or professional groups (unions, CUMA, GIEE, etc.). Intermediary actors therefore appear to play a key role not only in technology diffusion but also in information, training, testing, evaluation and development. This calls for a deeper exploration of the roles they play and how digitalization transforms these actors, taking into account – or not – the challenges of ecologization. This question is the subject of Chapter 6.

6

The Role of Cooperatives in the Development of Digital Technology

Interactions between digitalization and ecologization take place both at the level of the innovation system and individual practices. The latter can be seen as "embedded" in the innovation system (Cooke et al. 1997; Klerkx et al. 2012). This impacts them through a variety of economic, institutional, relational or cognitive mechanisms: regulatory constraints (public policies) or incentives linked to upstream (prices and access to technologies) and downstream (contracts, demand, etc.) markets, norms and representations conveyed, technical knowledge acquired through training or research, networks of relationships between actors, etc. Intermediaries can be defined as organizations or entities that establish links between two or more parties at a stage of an innovation process and participate in these mechanisms (Howells 2006). They link the different scales of the innovation system. In the agricultural sector, intermediaries are seen as central to innovation processes and are widely studied in the literature (Touzard et al. 2014). Several types of organizations can play the role of intermediary: producer organizations, cooperatives, private consultants, business networks, research transfer organizations, training, consulting organizations and even digital platforms (Klerkx and Leeuwis 2009a).

Agricultural cooperatives are a major intermediary in the agricultural sector. They represent a grouping of producers and thus participate in representing them in political and economic interactions, and in pooling resources. They also serve as a market interface, linking production and marketing, benefiting from technological and commercial economies of scale. Furthermore, they play a key role in knowledge transfer, implementation and support of public policies. They are particularly central to providing technical advice to farmers (Laurent et al. 2021) to such an extent that they have been the subject of specific debates regarding the separation of their sales and advisory functions on plant protection products (Chomel 2019). However, relatively few studies focus on cooperatives and their intermediary role in the innovation system (Filippi and Frey 2015). Agricultural intermediaries, and in particular cooperatives, have played a major role in the historical transformations of the agricultural sector and are being brought back to the forefront to face the new challenges of ecological transition, but also of the digitalization of agriculture.

The analytical chapters at the scale of the innovation system and at that of farmers' practices have highlighted the importance of cooperatives in disseminating technologies, training and supporting farmers, offering services based on digital technologies, and linking with downstream actors. It therefore seemed crucial to delve deeper into their role in digitalization and its relationship to ecologization. Indeed, cooperatives find themselves at the crossroads of knowledge transformations, public policies, access to technologies and also the monitoring (or sometimes control) of farmers' practices, both in terms of digitalization and ecologization. They are also linked to the various actors in the innovation system (research institutes, technical institutes, chambers of agriculture, etc.) and must meet the demands of downstream actors (agribusiness, large retailers, traders, etc.). This raises questions about how their role evolves in these two transformations and how they contribute to their integration. Our work raises several questions: How do agricultural cooperatives implement digitalization? Does this implementation depend on, or impact, activities related to the ecologization of members' practices? Do digitalization and ecologization change their raison d'être? Do they transform their organization and governance?

This chapter aims to understand the role of cooperative organizations in these transformations and in their articulation. It is based in particular on an article published in the *Revue internationale de l'économie sociale* (Recma) (Schnebelin et al. 2022b).

6.1. Cooperatives engaged in various forms of ecologization

Our investigations show that cooperatives have been engaged for about 20 years in a variety of actions, which respond to environmental challenges to varying degrees.

The ecologization of cooperatives and their members mainly concerns four areas of action (see Box 6.1): the reduction of inputs, soil management, landscape infrastructure and the reduction of greenhouse gas emissions.

1) The implementation of practices that reduce the use of inputs concerns herbicides, insecticides, fungicides or irrigation. These practices can be based on optimizing the application of inputs (adjusting doses, dates or methods of application), on substituting them with alternatives (biocontrol, mating disruption, mechanical weeding, eco-grazing, etc.) or through more global redesigns of practices (integrated pest management, intercropping, etc.).

2) Soil management has been identified as a major lever for the ecologization of agricultural practices, whether for plant health, nutrition or carbon sequestration. Soil management practices include tillage (or non-tillage), soil amendment, and weed management.

3) Landscape infrastructure is one of the practices implemented to address environmental challenges. Hedges, trees and low walls can serve as biodiversity reservoirs and provide ecosystem services.

4) We can also add practices aimed at limiting greenhouse gases or capturing carbon: for example, reducing soil tillage or reducing the passage of agricultural machinery.

Box 6.1. *Areas of action of cooperatives on ecologization*

These actions are organized according to different approaches. Some are linked to certifications (HVE[1], Terra Vitis[2], organic farming, etc.), which aim to promote these environmental actions to buyers. Other actions are linked to "internally conducted" projects, through experimentation, internal cooperative committees, or in collaboration with technicians. Other environmental actions are also carried out in relation to public policies, for example, the implementation of agri-environmental and climate measures, which allow access to Common Agricultural Policy (CAP)[3] subsidies, or participation in Economic and Environmental Interest Groups (GIEE)[4], also supported by the CAP. These approaches may or may not be the subject of communication and commercial promotion (cooperative website, wine back labels, events, etc.). Within viticulture, the terroir also becomes an object of ecologicalization and commercial enhancement, with, for example, two of the cooperative wineries having integrated into their communication strategy an animal emblematic of the local environment (lizard and shrike).

1 The High Environmental Value (HVE) certification is an environmental certification at the farm level, developed and managed by the State.

2 Terra Vitis is a certification based on environmental, social and economic criteria. It is specific to the wine industry. It was established by the profession and is recognized by the State.

3 *La politique agricole commune*, in French.

4 *Groupements d'intérêt économique et environnemental*, in French.

Environmental initiatives have become a stated priority for cooperative leaders, primarily to improve their image and meet new consumer expectations. However, opinions differ regarding their true economic value. Some focus on organic farming, while others emphasize the associated risks and production costs, as well as the difficulty of marketing these products. Less stringent certifications, such as HVE or sustainable viticulture, offer few immediate benefits but ensure compliance with regulatory requirements and buyer conditions, particularly for export. These certifications are seen as an important issue, supported by public policies such as the CAP.

With increasing societal pressure and demand for more environmentally friendly products, communication surrounding these initiatives is becoming crucial. These projects often involve the collaboration of numerous external actors, which sparks debate within cooperatives. Some members are highly committed, focusing on organic farming or agroecological innovations, while others are more hesitant, and content with the bare minimum required by regulations. This ecological transition thus creates a diversity of situations depending on members' personal backgrounds, age and activities. In this context, cooperatives play a central role. They must segment and add value to production, provide technical support to producers and train them, while building a strong brand image and acting as mediators with external partners.

Two main trends are observed: some cooperatives fully integrate environmental issues into their strategy, involving all their members in this dynamic, even going so far as to undertake ambitious collective projects (A and C cooperatives). Others, however, do not prioritize it, seeking to reconcile different models without a truly shared vision. These strategic choices are reflected in the rules for classifying and pricing grapes, with bonuses for organic or HVE-certified practices, for example.

6.2. Cooperatives involved in digitalization projects

Digitalization can be characterized by four main areas (see Box 6.2): vineyard management, harvest organization and plot selection, traceability, and communication. The cooperative wineries visited are all involved in some of these areas (see Table 6.1).

For cooperative managers and members, digitalization is primarily a response to administrative and logistical requirements, which have increased with wine segmentation, environmental regulations, private standards and export constraints. Digital tools are therefore designed to facilitate the work of employees, particularly during the harvest period, and to ensure product traceability. They also help manage the complex logistics of cooperative wineries, which are often spread across several sites. Digital technologies also offer an opportunity for cooperative wineries to reposition themselves in the field of agricultural advisory services, which is evolving with new challenges related to climate and the environment, certifications (such as technical support and audits) and regulatory changes (such as the separation between sales and advisory services, which limits the role of input suppliers).

Vineyard management for members: digital technologies include weather services, the (still limited) use of decision support tools (DSTs) for treatments, disease risk prediction, irrigation management and robotics (still in the prototype stage). DSTs involve using data, models and web or smartphone applications to predict situations (disease risk, pest risk, drought stress, fertilization needs, etc.) and help make agronomic decisions. These DSTs can be provided directly by weather station or probe providers, technical institutes (IFV), agrochemical companies or even input supply cooperatives.

Harvest organization and plot selection: for example, remote sensing services have been developed to assess vine vigor and grape ripeness, allowing for plot selection across the cooperative and determining harvest dates. Digital logistics tools are also used to plan the harvest and coordinate with members, manage grape deliveries to the receiving dock, and oversee the initial winemaking stages, which are more complex to organize with the expansion of wine ranges. These harvest organization and plot selection tools primarily aim to facilitate the work of cooperative employees during this intensive and strategic period for producing quality wines.

Traceability management: digital technology is also used to manage the traceability of the flows involved in the production and marketing of the cooperative's wine. This collection and organization of information is used to guarantee compliance with specifications (particularly HVE) or bulk contracts (export trading or brands). More broadly, it links, on the one hand, the winemaking and marketing structure and, on the other hand, the members (management of member accounts), but also the other stakeholders in the organization: suppliers, buyers, the ODG[5] and administration (purchasing and customer account management, inventory tracking and official quality labels). Cooperative wineries are thus committed to implementing digital traceability across all material and economic flows.

Internal/external communication: internally, the focus is primarily on informing and sharing knowledge about winemaking techniques, logistics, regulations and the cooperative's internal operations, including key moments such as the harvest season. Various communication channels are used: WhatsApp groups, SMS, intranet, email and videoconferencing. Some of these tools are not yet widespread (including those used for organizing board meetings, for example), but they are becoming essential for coordinating member relations, winemaking and sales activities, and cooperative management, especially when the cooperative operates across multiple sites. External communication is organized around supplier and customer networks, but it is also increasingly extending to consumers via websites and social media.

Box 6.2. *Areas of digitalization in which cooperatives are engaged*

5 An ODG is a defense and management organization set up on the initiative of a group of operators to ensure common production rules.

Cellar cooperative	Vineyard management	Harvest and plot divisions	Traceability	Communication
E	Treatment DST (2020) in testing	Parcel remote sensing (2019): under trial	Traceability by plot and at the cellar (2017); traceability of practices (2020)	Members' area on the website, internal blog (harvest organization) Website, online sales
F	–	–	Traceability of treatments (2018): free for winegrowers; diagnostic certification software; winery traceability software	Members' extranet area Website, online sales, social media
J	Weather stations with DST	Land parcel management software	Traceability at the winery, traceability certification (paid for by the winery)	Website, social media, online sales
M	Decision support tool for treatment, fertilization modulation test, autoguidance test, robot test	Harvest management, plot management software	Traceability in the cellar	Harvest organization application Website, online sales, social media
G	Weather stations with DST	Harvest management	Traceability in the cellar, traceability certification	Member area on the website Website, online sales
B	Weather stations with DST	–	Process traceability software	WhatsApp group for technicians and members Website, online sales, social media
C	Weather stations with decision support tools, irrigation DST	Land parcel management software	Process traceability software	Website, online sales, social media
HAS	Treatment with DST	Remote sensing of plots, harvest management	Process traceability software	Member area on the website, newsletter Website, online sales, social media

Table 6.1. *Development of digital tools by each of the cooperative wineries visited*

The implementation of these digital services relies on two types of collaboration: economic collaborations with companies specializing in software or digital technologies and partnership collaborations with agricultural organizations to offer training, conduct experiments and disseminate new technologies.

The cooperative plays a dual role in the digital sphere: it disseminates technological innovations to its members and invests in digital technology, sometimes deciding to charge members for these services. For example, some cooperatives have purchased and installed connected weather stations on their land. These stations collect data, which become accessible via a shared application for winegrowers and technicians. Cooperatives can also test and experiment with new technologies, even if they sometimes encounter difficulties in evaluating them. They also coordinate data collection, such as the water status of vineyards measured by probes, which is then used in decision-support tools. Thus, the cooperative takes on a new role by advising and supporting its members in the use of digital technology. This support can take the form of demonstrations organized by technology providers, or internal testing, thereby allowing the cooperative to take a more objective view of the technology providers' sales pitches.

6.3. New functions of cooperatives with digital technology

6.3.1. *Economic intermediation*

Cooperative wineries find themselves with new roles linked to their position as intermediaries.

First, as economic intermediaries, cooperatives establish relationships with new partners, such as companies that provide digital technologies, offer services or participate in environmental projects. Some cooperatives choose to collaborate with large companies to guarantee consistent service, while others prefer smaller structures for a more personalized relationship and tailored services (such as adding modules to software). Cooperatives also modify their existing relationships, for example, with suppliers of agricultural inputs, who may also offer technologies. Some cooperatives use weather stations and DSTs provided by input suppliers while others work with independent companies. Thus, cooperative wineries can become technology disseminators, offer related services or make collective investments. For software that ensures the traceability of treatments carried out by winegrowers, two strategies emerge: the first involves the cooperative covering the software costs, thus encouraging winegrowers to use it, which then simplifies traceability management for cooperative employees. The second strategy involves charging members for this service, thus giving them the choice of whether or not to invest in digitalization.

In this intermediary role, reconciling ecology and digitalization can be difficult. Indeed, the expectations and perceptions of partners vary for these two processes, and the economic and political interests of cooperatives can also come into play, raising questions about the neutrality of cooperatives (Yang et al. 2014).

In the grain sector, new economic relationships are emerging between cooperatives and agri-food industries, which are collaborating closely to establish specific sub-sectors with strong traceability. As in viticulture, crop cooperatives are encouraging the use of traceability software to meet the requirements of these industries, develop quality supply chains and access higher added-value markets.

However, software-based traceability appears less crucial in large-scale crops than in viticulture, where it is more important, even mandatory, due to segmentation requirements. Furthermore, it is less common and less readily accepted by cereal producers. Cooperatives acknowledge that the use of traceability software represents a major challenge, but they encounter technical and organizational difficulties, as illustrated by the example of a cooperative below:

> For our quality supply chains, we need extremely high traceability, from the field to the pack of pasta. I'm exaggerating, but only slightly; soon we'll have to put the member's photo on every grain of wheat, you see. In theory, we have all the tools to do all this, and to meet the demands from our agri-food industry partners, except that if the traceability tools are incomplete… so we're struggling today to gather all this information. (Manager of a cooperative in the Southwest)

From the producers' perspective, this traceability requirement is seen as intrusive, with the risk of "being monitored", the risk of dependency and of surveillance by the cooperative, as one farmer testifies.

> Everything must be transparent for the farmer, but nothing is transparent at the cooperative. We are increasingly tied to the cooperative. (Farmer 47)

However, the cooperative also appears as an interface that facilitates the implementation of digital technology, as evidenced by another farmer who "only goes through the cooperative to have the maximum number of bridges between [his] digital tools" (Farmer 83).

These quality-based supply chains in large-scale farming, which are developing within the context of an "economy of quality" encompass different quality conventions, which may emphasize digital technologies (industrial convention) or criticize them (Allaire 2002), just as in the wine sector. The information that aims to detail this quality itself becomes a quality that could be described as the informational quality of the product. This is illustrated by a testimony regarding the expectations of large retailers toward procurement organizations:

> For example, last week we were at a Smag meeting in Paris, and they brought in a mass retailing group who said, "We sell a product, that's true, but what we're missing is the story behind the product." And today we're well aware of that, but the farmer, since there's no compensation at the end, doesn't really care. (Southwest Trade)

6.3.2. *Technical intermediation*

Cooperative wineries play a key role in technical coordination to meet increasingly complex traceability requirements. They must combine different traceability software (for vineyards and wineries), often linked to modules related to certifications (such as VDD or HVE). This creates compatibility issues or complicates procedures. Cooperatives are therefore seeking to centralize this information to better manage their overall strategy.

Implementing these information systems requires support for employees and winegrowers in using the software, and sometimes even the direct handling of tasks for those who lack the necessary skills. This support is all the more important given that some winegrowers feel discouraged by the increasing number of online procedures, and that their digital skills vary.

To address these challenges, several cooperatives have covered the cost of software to support those "making the most progress", and/or they have implemented specific training programs to help "those experiencing difficulties with these tools". The main challenge is to develop an internal network of expertise capable of supporting this new role of technical coordination, leaning on digital tools.

6.3.3. *Knowledge intermediation*

Second, cooperative wineries are modifying their knowledge intermediation mechanisms, using new tools for knowledge transfer and exchange (DSTs, social networks, etc.). The challenge is to integrate these new knowledge tools with mechanisms that allow for the consideration of existing knowledge, the co-production of knowledge and the necessary learnings for ecological practices

(Chiffoleau 2005; Plumecocq et al. 2015). Cooperatives are finding themselves with new roles in building knowledge about digital technologies and ecological practices, and in selecting knowledge disseminated through digital tools. Indeed, the evaluation of innovative technologies or practices by the cooperative, as well as collective validation, appears necessary and decisive in the innovation process (Del Corso et al. 2015; Bechtet 2022). This again underlines the central role of knowledge in the digitalization process (Fielke et al. 2020).

As in viticulture, large-scale farming cooperatives are sometimes involved in technology assessment processes and report difficulties in carrying them out. Farmers are involved in these assessments, but some criticize the quality of the evaluations performed. One farmer testifies to having stopped using a service offered by the cooperative.

> Let's say we were part of the pilot group to test it. We were told it would be tested for two or three years. The first year we tested it, on two plots, it didn't produce the expected results. However, before we improved the techniques, they marketed it to all the other members, using us as a bit of a front. I wasn't the only one; there were about ten of us. For me, it wasn't sufficiently tested and validated before being marketed. And then they also charge €20-25/ha. (Farmer 47)

Indeed, conflicts of interest can arise between the cooperative's role in evaluating digital technology and its function as a service provider. The cooperatives encountered in large-scale farming tend to focus on offering new paid services to farmers, rather than on collective investment, as is sometimes the case with cooperative wineries. This could be explained by the fact that large-scale farming cooperatives are seeking new profit margins in a context of declining input sales, whereas cooperative wineries do not sell inputs.

In both large-scale crop farming and viticulture, cooperatives must address the need for upskilling their technicians to support these new services. In large-scale crop farming, technical advice has long been a core mission of cooperatives, and digital tools are introducing new areas of expertise or complementing existing advice. In viticulture, digital tools are supporting the development of new technical advisory services, in a context of evolving member needs (particularly for certification) and a decline in the availability of consulting services. Some arable farming cooperatives are developing dedicated departments and specialized consulting services for the implementation and use of digital technologies. Others are relying on existing services and advisors.

6.3.4. *Regulatory intermediation*

Finally, cooperatives can link these two processes via their role as political and regulatory intermediaries. On the one hand, they support producers in complying with environmental regulations, an area in which traceability software and digital communication tools are used. Furthermore, cooperatives are organizations that can be tasked with implementing public policies (Plumecocq et al. 2015). They end up as drivers and partners in ecologization and/or digitalization projects, with public support that they must coordinate according to their internal resources and objectives, primarily focused on value creation. Furthermore, cooperatives, as marketing organizations, face significant challenges in integrating private regulations, challenges that are reinforced by both ecologization and digitalization.

Traceability management (from practices to marketing) thus becomes a major function of cooperatives, since it is a condition for access to markets, segmentation and value creation, and involves training and support for winegrowers in taking control of traceability software.

6.3.5. *The evolution of their role as a collective organization*

But cooperatives are not impacted by digitalization and ecologization only because they are intermediaries with respect to their members. They are also innovative in their own investments as well as their productive and commercial activities, adopt technological and digital innovations, can bring about new ones and participate in changing the institutions and regulations of the sector (Yang et al. 2014). The interviews show that, particularly regarding information and communication, all cooperatives use digital technologies. Internally, cooperatives use newsletters and even set up exchange networks on social media. Exchange networks between producers are also being built beyond the cooperative in smaller groups or groups more focused on objects of common interest.

6.4. The role of bridging digital and environmental challenges

6.4.1. *Articulation of these issues through the functions of intermediaries*

As economic intermediaries, through their purchases and investments in equipment, software and services, cooperatives seek options that can integrate both environmental and digital concerns, potentially leading to cost savings. However, digital technologies and environmental issues are not always aligned, and they must make trade-offs that often lead them to prioritize either ecologization or digitalization, choices which are made according to their business strategies.

The use of digital technology to communicate about ecologization initiatives opens up new perspectives. For example, some cooperatives we met with communicate digitally about member initiatives or assert an environmental identity for the cooperative.

As knowledge intermediaries, faced with the multiplicity of information which is sometimes poorly adapted to the local context, the members and employees surveyed emphasize the importance of the cooperative to collect and capitalize on data with regard to experiments and more ecological practices, share training content and "sort" between information carried by different organizations in the sector.

Furthermore, in some cases, the development of digital services is also conceived as a way for farmers to build knowledge about their soils, with the aim of initiating comprehensive discussions on plant health, fertility and irrigation. These cooperatives hope to leverage all the data collected to analyze practices, identifying practices that work well but are limited by the time and resources available to conduct such analyses.

In the cooperatives studied, the link between these two processes remains limited, even though synergies are emerging via the development of traceability and digital communication. Digital technologies are indeed being implemented to meet a variety of objectives without necessarily integrating ecologization. The capacity of these technologies to address the challenges of ecologization depends on the evolution of public policies, public and/or private R&D strategies, and the technological offering, but also on actions undertaken by the cooperatives themselves. These cooperatives play an intermediary role within the sector for both ecologization and digitalization, and they can therefore adjust, harmonize or even design innovations that are part of both processes. With digitalization and ecologization, cooperative wineries could thus reaffirm their purpose and missions, and allow winegrowers to better control the innovations and transformations of their activities. This perspective seems to be shared by all wine companies and all agricultural cooperatives (Ciruela-Lorenzo et al. 2020; Candemir et al. 2021), calling for the expansion of our research to other productions, regions or countries. By repositioning themselves as organizations that promote local coordination and learning for ecological and digital innovations, cooperatives can thus renew their project and be at the heart of the major new transformations in agriculture.

6.4.2. *Differences according to the type of cooperative*

Digitalization in large-scale farming cooperatives appears to be a more individualized process, involving individual services, advice and training, whereas in viticulture, the approach is more collective and often managed by cooperative wineries (investment, support). The differences observed between these sectors can be linked to several factors. First, the fact that, on the one hand, we are dealing with basic commodities and, on the other hand, with products (wines) whose quality is linked to the collective management of territorial resources, considered as forms of public goods (terroir, landscape, place name, etc.), necessitates greater collective action in the implementation of innovations that could modify the intrinsic (e.g. grape ripeness) or extrinsic (image) quality of the product. It should also be emphasized that cooperative structures in large-scale crop farming are much larger than those in viticulture, both in terms of the number of members, the land area involved and the number of employees. This can affect their needs, but also their governance and the management of internal heterogeneities, with members who know each other less well and work in more varied contexts. Furthermore, at the individual level, the use of digital technologies, particularly for production, is more limited in the wine sector which was studied.

Several hypotheses can be put forward to explain these differences. First, vineyards are smaller, which may also explain the greater reliance on collective approaches regarding digitalization. Second, winegrowers reported that some digital technologies are not suited to their plots, which are highly fragmented, with many slopes and obstacles. Third, the optimization of inputs, which remains central to most currently available digital technologies, is undoubtedly less crucial in viticulture, at least for fertilization. Indeed, the proportion of costs allocated to inputs is lower in viticulture than in large-scale crop farming. Furthermore, maximizing yield is not often a goal for winegrowers, who tend to aim for maximizing quality, which may imply limiting yields (specifications impose a maximum yield).

6.5. Contributions and perspectives

Focusing on cooperative wineries, and then more broadly on agricultural cooperatives, provides a complementary perspective to the findings of Chapters 4 and 5. First, the role of intermediary organizations, in this case cooperatives, is confirmed in the digitalization of agriculture. As discussed in previous chapters, they disseminate technologies, can participate in their evaluation and support and train winegrowers or (to a lesser extent, depending on the sector) farmers. They can also invest directly in digital technologies, allowing members to benefit from economies of scale or to assume some of the risk associated with implementing these innovations, which are not always suited to local conditions or skills. They also serve as a point of contact for digital organizations, enabling them to articulate their needs. Finally, they play a role in linking digitalization with other challenges, particularly environmental sustainability. This coordination can be achieved, in particular, through experimentation, by aligning regulatory and technological developments, and by the collective building of knowledge and skills. However, this coordination is constrained by developments which are external to cooperatives, but also by their internal functions. Indeed, cooperatives are not neutral organizations; they have their own interests, which can sometimes conflict with those of some of their members or with societal objectives (Yang et al. 2014).

This work also highlights and questions the unique characteristics of cooperatives compared to other intermediary organizations. As a system of interaction between actors coordinated by a specific governance structure, but also by their role in political representation, cooperatives can engage differently from other intermediary organizations. Furthermore, the two innovation processes studied involve the work and skills of cooperative employees and can mobilize local resources, the use of which encourages collective action for which the cooperative can provide a privileged framework. This calls for further research on the transformations of labor relations and collective action within cooperatives.

7

Discussion: Digital Technologies for What Kind of Ecological Agriculture?

The three preceding chapters examined how digitalization is implemented at different scales within the agricultural sector and how it can be integrated into various pathways to ecologization. At the innovation system level, we demonstrated that expectations and perceived risks differ depending on whether stakeholders are involved in organic or conventional agriculture. They do not implement the same digitalization strategies. However, these differences are often overlooked by digital actors. This creates a risk that technologies will be developed solely according to the framework of conventional agriculture. At the level of large-scale crop farms, the usage patterns of digital technologies vary according to the agricultural production model and the ecologization trajectories. The results confirm the hypothesis that current digital development tends to reinforce trajectories of expansion, specialization, growth in salaried employment and subcontracting, with limited interaction with the ecologization of the sector. Our results further show that cooperatives are central to the processes of ecologization and digitalization. However, they encounter difficulties in coordinating these processes due to objectives, partners and public policies that are not necessarily aligned. These cooperatives nevertheless have room to maneuver to reaffirm their role and enable winegrowers to exert a counterweight to the digitalization trajectory. This discussion chapter aims to draw lessons from the analyses presented in this book.

In this discussion chapter, we offer a cross-sectional analysis to clarify how digitalization interacts with the ecologization trajectories of agriculture in France. Our results show that a diversity of digitalization processes are at work, linked to production models and associated with transformations in the innovation system. Digitalization is part of longer term transformations in the agricultural sector and, so far, has tended to reinforce pre-existing dynamics rather than revolutionize the sector's organization. Digitalization appears to be in opposition to the ecologization of the sector, even if some hybridizations are occurring or seem possible.

7.1. Digital trajectories linked to agricultural models

7.1.1. *The diversity of digital trajectories*

"What digital technology will allow, through platforms, is to multiply opportunities for 'working together' in a profession that already has the capacity to collaborate", announces the CoFarming association in its white paper published in 2017[1], building on the announcement made in a 2016 publication by the Ministère de l'Agriculture et de l'Alimentation (2016, p. 12), Alim'Agri: "The Big Data revolution implies pooling resources: everyone is learning to cooperate!" The Ministry adds that "for this Big Data to serve the farmer, data must be made accessible to the greatest number of people in an open innovation approach that fosters the emergence of startups and new operators" (ibid., p. 10). On the other hand, industry representatives are asking: "Is France lagging behind in digital agriculture[2]?", while the authors of the book *Agricultural Robotics* distinguish between farmers based on their innovative behavior: "Not every farmer, however, is meant to be an early adopter. There will always be people who prefer to employ old-school practices and traditional methodologies" (Lenain et al. 2021, p. 16).

Discourses on digital technology in agriculture thus oscillate between visions of "open innovation", "co-construction," or "participation", and a linear, top-down conception of innovation, which the concepts of "acceptability" of technologies, "resistance to change", "laggards", or "pioneers" refer to – concepts widely used by digital companies and in institutional documents. The empirical work presented in this book highlights the nonlinear and heterogeneous nature of digital development, linked to the diversity of agricultural models.

Whether at the level of individual farms, agricultural cooperatives, or the innovation system, our results show that there are multiple digitalization trajectories. For example, some farmers use numerous digital technologies with their agricultural machinery without ever going online or using social media to find information. Others, on the contrary, are very active in forums and have websites, but have no digital devices on their agricultural equipment. At the level of cooperatives and

1 See CoFarming (2017, p. 26).

2 Fontaine, S. (2017). Indice mondial de l'innovation 2017 : la France en retard sur l'agriculture numérique ? [Online]. Available at: https://www.industrie-techno.com/article/indice-mondial-de-l-innovation-2017-la-france-en-retard-sur-l-agriculturenumerique.50218.

other agricultural organizations, different strategies are also at play: the marketing of digital production services in partnership with digital companies, developing digital platforms or various software programs in-house, collectively investing in technology, renewing information and communication methods via digital means, and so on. Actors make different choices concerning digitalization, in terms of the types of technologies developed or mobilized and their uses, but also in terms of economic models (collective investment, occasional use of a digital service, regular subscription and in-house development) and in terms of partnerships (purchase, training, experimentation, service provision, etc.).

Our results also show that digitalization trajectories are not linear. Fieldwork reveals phenomena of adjustment, trial and error, abandonment and occasional use, both at the farm level and at the organizational level. Some farmers try a technology, stop using it and sometimes return to it. Others use a technology occasionally through a neighbor or a service provider. Some users abandon a technology because they cannot use it correctly, but also because they have not found it beneficial, because they change their farming practices or revise their economic strategies and decide to limit investments. These observations reinforce and complement the literature on "droppers" or "dis-adopters" – that is, farmers who stop using a technology – a body of literature that highlights the lack of advisory services or the observed lack of benefit as a factor in abandonment (Sutherland et al. 2022). The abandonment of a technology can come from a decision following a self-assessment or from a more global strategic bifurcation, constituting a form of "innovation by withdrawal" (Goulet and Vinck 2012).

We also showed that some cooperatives try out and test technologies, and then choose whether or not to promote them to their members. Others regularly change providers for digital services. The results reveal learning processes, for example, with cooperatives or farmers who initially use a digital service (such as field remote sensing), and then use the data produced and the knowledge acquired for several years without reusing the service. Some adapt their use of software or technologies to better meet their needs, as highlighted by Higgins et al. (2017). Others adopt a digital technology in response to a standard, a regulatory change or a contract. Furthermore, the description of the technologies and their uses shows that the technologies are interdependent, a fact that is generally under-researched in studies on adoption (Barnes et al. 2019; Lowenberg-DeBoer and Erickson 2019).

These interdependencies can create threshold and reinforcement phenomena (Clapp and Ruder 2020): once you have one technology, it is easier to access others (e.g. the use of input variable-rate, or section control technologies, is contingent upon having GPS guidance. They can also generate phenomena of exclusion: if access to one technology is difficult, this limits access to a range of other technologies. These contingencies and technological interrelationships that influence the technological trajectory underline path dependence mechanisms highlighted in evolutionary economics research (David 1985).

These observations suggest moving beyond an overly simplistic and linear view of innovation adoption, such as the one initially proposed by Rogers (2010), which distinguishes between "innovators" and "laggards". It seems more relevant to consider the conditions under which using a technology becomes worthwhile. Indeed, our results show that heterogeneity at the level of farms, as well as within sector organizations, is not solely explained by individual characteristics, but rather arises from a diversity of production models associated with different objectives, technical constraints, perceptions, knowledge and available resources.

7.1.2. *Aspirations associated with agricultural models*

At the level of agricultural organizations and cooperatives, the development of digital technology depends on the sectors involved, but also on the type of ecologization these organizations are implementing. These forms of ecologization notably impact their expectations regarding digital technology. For example, the expectations of organizations linked to conventional agriculture are primarily focused on improving profitability, traceability to enhance and segment markets, developing new commercial services, enhancing the image and attractiveness of agriculture, improving working conditions, optimizing practices and reducing environmental impacts. Some of these issues reflect a vision of so-called "weak" ecologization. Organizations linked to organic farming emphasize their expectations for digital technologies, which are accessible to all, promoting the exploration, analysis or systemic design of agroecosystems, strengthening horizontal exchanges, and enabling the promotion of products, services and knowledge. They also have expectations for technologies that could solve problems specific to organic production, notably facilitating mechanical weeding and soil cultivation (to combine organic production and soil conservation, in particular), or limiting the use of inputs, especially copper in organic viticulture. In viticulture, expectations for digital technologies may converge more around issues concerning the reduction of inputs, which is common to several approaches to ecologization (conventional and organic viticulture). The issues highlighted by organizations associated with organic agriculture then relate to the different forms of ecologization (optimization and redesign), but also to a socioeconomic model.

Alongside divergent expectations, the perceived risks of digitalization also differ. Actors within the conventional paradigm fear the appropriation of data and value by digital companies. They are concerned about technical problems, such as technology compatibility, bandwidth issues and reliability problems. Organic farming actors, on the other hand, mention greater risks of dependence, loss of power and knowledge, as well as risks of standardization or exclusion. These findings notably contributed to the chapter on the risks of digitalization in agriculture in the INRAE-INRIA white paper (Bellon Maurel et al. 2022). These differences in representations and perceptions linked to the agricultural model partly explain the differences in strategies and implementation of digitalization by agricultural organizations. They reveal a potential diversity of digitalization trajectories.

7.1.3. *Uses embedded in production models*

At the farm level, the literature highlights the impact of factors such as farm size in studies on the adoption of precision agriculture technologies (Barnes et al. 2019; Konrad et al. 2019). We show that digital uses depend more broadly on the farm's production model (level of capitalization and nature of technologies, form and organization of labor, product quality and ecologization) in interaction with its integration into the agricultural innovation system (advice, marketing, professional networks). For example, the significant use of digital technologies for production (DTPs) is linked to farms' expansion, but also to their integration into the advisory and marketing networks of seed companies or cooperatives. These actors can indeed encourage or impose the use of digital technologies. The implementation of organic farming or diversified production is less strongly associated with the use of DTP, with the exception of large-scale mixed organic-conventional farms. The use of digital technologies for information and communication (DTC) is becoming more widespread, particularly among farmers already involved in experience-sharing groups. The literature highlights the role of cooperatives in the adoption of digital technologies (Barnes et al. 2019). In this work, we demonstrate the role of production models, which can be implemented by cooperatives, in the adoption and use of these technologies.

Our findings on the digitalization of agriculture thus confirm the systemic, interactive and collective nature of the innovation process, linked to techno-economic, individual and politico-institutional determinants, as well as to local professional groups (Darré et al. 1989; Ansaloni and Fouilleux 2006). They encourage us to reconsider the individual and linear view of the innovation process embedded in economic and political representations of innovation. The mobilization of this individual conception of innovation naturalizes and individualizes the deployment of technologies. It serves as a political and economic justification for supporting businesses, in order to "catch up" (Martin 2019), and masks the diversity of needs, expectations and technological logics in agriculture. Digitalization trajectories depend on economic and agronomic models, values and objectives (Klerkx and Begemann 2020). They therefore depend on the structural changes in the agricultural sector while also causing it to change.

Digitalization encompasses a diverse range of new technologies, services and actors in the agricultural sector. Some of these technologies, such as geolocation or model-based decision support tools, originate from research, followed by the industrialization of these inventions, their dissemination through intermediary organizations and their adoption on farms to meet the needs arising from their industrialization. A portion of digital technological development thus occurs from a diffusionist perspective, which does little to integrate the diversity of needs during the design process (Bronson 2019). Innovation policies, most often supply-side focused, generally propose partnerships between research and businesses and appear to be based on a linear conception of innovation, according to our field observations. Agricultural organizations then play a "classic" intermediary role in dissemination, but also serve as spaces for training and experimentation. Several organizations surveyed, particularly those in organic farming and wine cooperatives, emphasize the importance of training and support for farmers (Newton et al. 2020). Agricultural organizations can also be partners in digital projects or initiators of innovation. While we observe some forms of structuring of actors around digital technology in agriculture, these remain largely organized around the supply side (research/business consortia, startup associations partnerships between digital companies and agricultural businesses) and involve farmers very little. There are, however, some partnerships being established between digital actors and agricultural organizations, but these tend to focus on the dominant agricultural model, without, until now, taking into account the heterogeneity of the agricultural sector.

Digital technology in agriculture thus appears as a set of innovations with specific characteristics. Far from being niche innovations, digital technologies affect a large portion of agricultural organizations and farms. Their development is not solely exogenous; it results from partnerships between digital and agricultural actors, as well as from projects internal to agricultural organizations. Digitalization is linked to a set of arrangements between actors, who participate in this process in heterogeneous ways according to their interests, vision and power, and to public policies, research and development, as well as economic transformation. These digitalization trajectories, although specific, are part of longer term dynamics, which we will discuss in section 7.2.

7.2. The continuation of historical transformations of the agricultural sector

The digitalization of the agricultural sector is sometimes presented as a "new agricultural revolution" (Barrett and Rose 2020; Rolandi et al. 2021) and sometimes, conversely, as a continuation, or even a reinforcement and acceleration, of the industrial transformations of agriculture (Bronson and Knezevic 2016). This section contributes to this debate. The industrialization of agriculture can be characterized by the specialization, intensification and enlargement of farms; the standardization of production; the increase in wage labor; the growing use of energy and materials from outside farms; as well as by increasing integration into value chains (Diry 1988; Daviron 2021). Our results tend to support the idea that digitalization is developing as a continuation of the historical transformations of agricultural production models and their industrialization, while integrating new aspects and even opening up the prospect of new trajectories.

7.2.1. *Digital technology as an accelerator of ongoing transformations*

At the level of production models, characterized by their productive organization (methods, technologies, resources and spatial organization) and biotechnical organization (agricultural practices and relationship to nature), forms and organization of labor, quality of products and their market integration, the results of Chapter 5 show continuities, but also paradoxical effects in the transformation of organizations, agricultural practices and labor.

7.2.1.1. *Expansion and specialization*

Fieldwork conducted on crop farms shows that some of the digital technologies used promote specialization, intensification and farm expansion – dynamics already underway since the mid-20th century (Servolin 1989; Allaire and Boyer 1995). Indeed, some technologies, particularly decision-support tools, only function with certain crops, and therefore may encourage an increase in the land area dedicated to them to make the investments profitable. It can also be assumed that investment in digital irrigation technologies tends to boost the production of irrigated crops. Technologies based on a single type of production thus lead to the pursuit of economies of scale and technological orientations that favor specialization trajectories. The objectives of intensification and increased production are integrated into the design of digital technologies (Bronson 2019), and the farms that make the most use of digital technology are generally those oriented towards greater productivism. The use of digital technologies also facilitates farm expansion by increasing labor productivity or facilitating labor management and control. These technologies also seem to offer new possibilities for the substitution of inputs and processes, which was also observed with the motorization of agriculture in the second half of the 20th century (Byé et al. 1989).

7.2.1.2. *Standardization*

The effects on standardization are more ambivalent. On the one hand, digital technologies enable the individualization of advice (Stone 2022), taking into account soil heterogeneities and climatic conditions. On the other hand, they are also perceived by some actors as forms of standardization of technical approaches and advice. Our surveys show that practices that do not conform to these standards can create problems when using technologies that are not configured for unconventional inputs or techniques. For example, farmers encounter problems using variable-rate application with organic fertilizers, or with data entry in software when using seeds or inputs that are not listed in the software. Traceability standards, particularly those linked to contracts and specifications from buyers, also seem to lead to a standardization of practices.

But this traceability also offers new possibilities for segmenting production, and therefore for promoting diversity, widely used in viticulture, but also applied to cereals with the development of quality supply chains (Heintz 1990). Furthermore, the use of DTPs is associated with new production methods, such as mixed farms or farms practicing conservation agriculture. The use of DTCs also sometimes leads to changes in practices, particularly regarding cover crops and conservation agriculture. Digital technology can therefore promote standardization within segmentations, which are part of an economy of quality (Nicolas and Valceschini 1995; Allaire 2002). Rather than standardizing toward homogeneous production, digital technology standardizes toward a segmentation of production. The effects of digital technology on the standardization of production methods are still little analyzed in the literature: they are sometimes mentioned as a hypothesis (Prause et al. 2020; Fraser 2021), but to our knowledge there are no empirical studies that delve deeper into this aspect.

7.2.1.3. *Decisions and knowledge*

Digitalization is also integrated into the transformation of decision-making and knowledge management methods within farms (Rose et al. 2018). These new technical artifacts can be seen as a form of outsourcing of decisions and responsibilities, associated with a transfer of decision-making skills to firms (Carbonell 2016; Carolan 2017b), or as a form of automation of intellectual labor (Lele and Goswami 2017). Indeed, the use of certain digital technologies partially modifies the decision-making process by adding new data, new levers, models or algorithms and new actors (technology providers). They also modify the process of managing data and knowledge on the farm: recording historical practices, data processing on the computer, new forms of data visualization, etc. Our results are paradoxical here as well. They highlight the perceived risks of a loss of decision-making autonomy, associated with a loss of power, which confirms findings in the literature (Jeanneaux 2018). Decision-making processes may include forms of skills transfer to firms (Fraser 2019). The construction and processing of data appear opaque and poorly understood by users, who sometimes question its validity or relevance.

However, our observations tend to put the importance of these skills transfers and outsourcing into perspective. Indeed, many users report a certain degree of skepticism regarding the knowledge derived from digital decision support tools (DSTs) or the internet. The advice and knowledge transmitted through digital technologies are often, when the farmer has the opportunity, contextualized and put into perspective with local knowledge, personal experience and advice from their professional network, which corresponds to the "re-scripting" of the DST described by Rose et al. (2018). They are also adapted to the objectives and risks that farmers are willing to take.

Some farmers, however, do not have this option. Indeed, several technologies that integrate sensors, modeling and automated applications prevent any user intervention. Sometimes, it is the arrangements associated with these technologies that prevent the user from intervening. For example, one farmer testifies that he is contractually obligated to apply the fertilizer dose prescribed by the variable-rate application map sent by the seed company that buys his crops.

7.2.1.4. *Work and outsourcing*

The development of digital technology also appears to be a driver of transformations in agricultural labor. The literature on digital technology in agriculture, when it focuses on labor, does so primarily from an individual perspective. Studies show how digital technology modifies professional identity and the nature of work, and also impacts meaning and job satisfaction, negatively in some studies and positively in others (Bronson 2019; Hansen and Stræte 2020; Rose et al. 2021). Our surveys reveal a similar ambivalence in the effects of digitalization on individuals. On the one hand, digital technologies save time on certain tasks and make them easier, improve knowledge, or reduce the arduousness of work (a frequently cited example is GPS-assisted driving, which prevents the tractor driver from having to regularly turn around). On the other hand, they come with new tasks that can offset the time saved, give a sense of control, and reduce autonomy (particularly for equipment repair), involve tedious and less appreciated tasks and harm relationships. Carolan (2020) defines these effects of digital technology on work as micropolitical and emphasizes that they are essential to understand and take into account. In our research, these micropolitical effects are also ambiguous.

The use of digital technologies on farms is also linked to broader transformations in the organization of work within production models, the study of which helps us understand contemporary developments in agriculture. While agricultural modernization was essentially based on the family farm model, digital technologies appear to meet the needs of the growth of salaried employment and subcontracting in agriculture within models that remain family-run, but whose capital intensity brings them closer to "corporate agriculture" (Hervieu and Purseigle 2009; Nguyen and Purseigle 2012). Indeed, the use of DTPs is associated with farms that rely more heavily on salaried labor and/or agricultural service activities. It should be noted that the increased use of hired labor alongside a decrease in family labor is a general trend in the agricultural sector, as is the rise in subcontracting (see RGA 2020[3]). In our surveys, digital technologies appear as a means of organizing, recording and controlling work (both one's own work and that of employees) on farms and in agricultural contracting companies. These results support the hypotheses that digital technology in agriculture promotes forms of surveillance and control of employees' work on farms (Rotz et al. 2019b; Giles and Stead 2021; Prause 2021). More broadly, our work shows that digital technology serves as a means of recording, tracking and controlling work (services, partners and employees) within a context of farm expansion and increasing complexity in farm management.

Providing agricultural services can create a need for digital technologies and be a way to make investments in digital technology profitable. This leads to forms of expansion through service provision, corresponding to an increase in cultivated areas without increasing the farm's overall size. In the farms examined here, cultivated areas are sometimes multiplied tenfold thanks to these services. The service can be focused on a function related to a specific machine, such as harvesting or sowing a crop. It can also be a turnkey service, encompassing all technical, and even administrative and decision-making, operations. Conversely, using services is a way for farmers to limit capital investment, mitigate risks and thus better manage their financial situation (Nguyen et al. 2020). It also appears to be a way to access technical and economic information, with service providers then playing the role of new prescribers. A parallel can be drawn between this increasing use of service provision and the trends toward greater flexibility in work associated with the development of digital technology in society (Srnicek 2018), but also with forms of invisibility (Gruszka and Böhm 2020), as the work of service providers is much less documented in the literature and in public statistics.

In society, digital practices are sometimes associated with a blurring of the lines between professional and personal life (Casilli 2019). Agriculture is already a type of work where the boundaries between these spheres are very blurred (Gregoire 2002). The literature has highlighted the fact that digital demands (alerts, notifications) can increase this blurring and work-related stress, particularly in livestock farming (Hostiou et al. 2017). In our surveys, digital practices for information and communication also appear to lie at the boundary between work and leisure, with farmers reporting that they go on social media "out of curiosity", "just to see" or because they "like it". This blurring is also found among agri-YouTubers, farmers who showcase themselves in their work, but also in their personal lives,

3 See: https://agriculture.gouv.fr/recensement-agricole-2020-julien-denormandie-presente-les-premiers-resultats-du-recensement-decennal.

through regularly published videos (Rénier et al. 2022). On the other hand, the effects of permanent digital solicitations have been little discussed, with the exception of advertising solicitations on the internet, which underlines the difference in issues between plant and animal production, where issues of work-related constraint remain more pronounced.

7.2.2. *Restructuring of the agricultural innovation system*

The development of digital technology is integrating and transforming the production models of agricultural operations. The transformations of their productive and biotechnological organization, their organization of work, their type of production and their market integration are linked to broader sectoral transformations (Boyer and Freyssenet 2000), while also showing forms of continuity with past sectoral transformations.

7.2.2.1. *Role of the State and of industries in innovation*

Regarding the emergence of technologies, research continues to play an important role, particularly in geolocation, modeling and precision agriculture technologies (Oui 2021b). At the upstream firm level, our results suggest a continued shift from product to service, a trend that began with industrialization and contributes to the restructuring of the agricultural input supply sector, notably through mergers and acquisitions (Byé et al. 1989). However, this development is driven by new extra-sectoral dynamics, through projects with companies such as Airbus, Orange and Microsoft. Our results also show other forms of emerging digital innovation, some of which, such as software, platforms or communication tools, are designed within agricultural organizations, by and for themselves and/or farmers.

However, some adoption dynamics stem from a power dynamic favoring downstream industries (agribusiness, trading, distribution, etc.) that impose innovations on farmers (Meynard et al. 2017). This observation primarily concerned production processes (seeds, fertilization, etc.). Our results show that organizational and digital innovations are now also at play. For example, contracts for the production of quality wheat for the agribusiness industry impose the use of digital fertilizer application technologies. Export constraints, specifications from large retailers and specific contracts require or encourage the use of traceability software. These innovations may be associated with final objectives, which allow for some flexibility in decision-making regarding how to achieve a given goal, or with means-oriented objectives, which impose the use of the technology in question (Cholez et al. 2017). The dynamics of digitalization thus appear to partly extend forms of integration of farms into upstream and downstream agro-industrial or commercial structures (Bartoli and Boulet 1989), particularly via data.

Digitalization also raises questions about the role of the state. During the modernization period, the state helped guide and pace technological progress in agriculture, but also participated in its genesis (Byé et al. 1989). Indeed, technological innovation is not solely the product of private dynamics; it is often associated with significant state support, including in the digital sector (Mazzucato 2015). Agricultural policies, in particular, incorporate a series of incentives, constraints and standards that guide the transformation of models and innovation, such as land policy, tax policy, training support and even policies for market stabilization and risk management. In France, agricultural policies are influenced by the persistence of neo-mercantile interests – that is, interests seeking to modernize agriculture and support exports – through instruments that support farm competitiveness, such as investment support (Trouvé 2009). This support can then encourage farm investment in digital technologies. Our results show that agricultural policies influence digital usage through their requirements (traceability, accounting) and the administrative procedures they entail (online reporting, for example) (Mesnel 2017). Digital traceability also meets the needs of supply chains in a context of deregulation by states, as already demonstrated in studies conducted in the United States (Wolf and Buttel 1996; Wolf and Wood 1997).

However, the development of digital agricultural technologies is also supported by supply-side innovation policies. The state plays a role in guiding technological developments, but more indirectly, via the funding of startups, public–private partnerships, and research projects. These innovation policies are primarily economic policies, the arbitration of which falls beyond the remit of the Ministry of Agriculture, as highlighted in an interview by the head of the Ministry of Agriculture's digital affairs department. The development of digital technologies and the role the state plays in them could contribute to a reorientation of innovation policies toward a more neoliberal perspective, a perspective that already appears dominant in other countries regarding agricultural digitalization policies (Fraser 2019; Stone 2022) (specifically concerning Ireland and Southern European countries). Our results show that these innovation policies could then contribute to forms of exclusion of certain actors from the agricultural sector, who do not envision the same economic models for digitalization or who do not constitute an attractive market for these companies. Furthermore, these policies give private companies an increasing role in guiding innovation, hence the importance of firms such as John Deere or Bayer in the trajectory of international digital development (Bronson and Knezevic 2016; Prause et al. 2020). "Corporate Innovation Systems" (Lundvall and Rikap 2022) can then interact with sectoral innovation systems in financing, structuring actors and knowledge, training users, developing digital technologies and organizing the sector, thus contributing to the construction of technological trajectories.

This is all the more important given that digitalization leads to forms of intellectual monopolization and firm concentration (Rikap 2021). Indeed, current transformations of capitalism are notably linked to forms of data extraction and accumulation, driven by the growth and competition of firms in a capitalist economy, and facilitated by platform organization, which would lead to platform capitalism (Srnicek 2018). The platform economy encompasses "economic activities where tangible or intangible resources are exchanged between providers and users by the way of centralized electronic platforms" (Montalban et al. 2019, p. 3). The expansion of platform capitalism could then lead to a restructuring of the agricultural sector with new forms of concentration and monopolization (Giles and Stead 2021).

Based on an analysis of agricultural policies in diverse countries, Trouvé (2009) identifies different forms of agricultural policy, which correspond to different compromises between divergent interests stemming from power relations specific to each context. These interests are characterized by three ideal types:

– neo-mercantile interests, which aim to increase production and the trade balance of the agricultural sector, through measures to support exports, protect markets and support the modernization of farms;

– neoliberal interests, which aim to promote the internationalization of trade and increasing commercialization of activities;

– multifunctional interests, which aim to support the environmental and social functions of agriculture.

These questions are a major current focus of research and policy in the field, addressed, for example, in the PADAC[4] project, a research project on "Digital Platforms in the Agricultural Sector and Dynamics of Collective Action"[5]. They echo debates on innovation strategies to adopt: supporting R&D or supporting the deployment of technologies through demand.

Box 7.1. *The different forms of agricultural policies*

These lines of thought raise questions about the role of digitalization in the evolution of agricultural policies. It could strengthen the hegemony of neoliberal interests to the detriment of neo-mercantile and multifunctional interests.

---

4 See: https://anr.fr/Projet-ANR-20-CE26-0003.

5 Plateformes numériques dans le secteur agricole et dynamiques de l'action collective, in French.

7.2.2.2. *The circulation of knowledge*

The development of digital technology also raises questions about the circulation of knowledge within the sector. On the one hand, digitalization is structurally altering the organization of consulting services with the arrival of new actors and digital technologies. Data, created by digital technologies, can be transformed into predictive capabilities and feed into consulting technologies or tools that can be sold. Consequently, agribusiness companies are seeking to accumulate data, a phenomenon described by the concept of data grabbing (Fraser 2019). In contemporary capitalism, this logic of accumulation, based on personal experience data transformed into predictive capabilities and influence that can be resold, is called "surveillance capitalism" (Zuboff 2019). The development of digital technology would thus contribute to a logic of accumulation and expansion of agribusiness companies toward a "surveillance agriculture" (Rose et al. 2021; Stone 2022). These companies would use data to create forms of individualization and commercialization in a logic of reinforcing the privatization of agricultural advisory services, a process that already began several years before (Labarthe 2009). Our results show that many agribusiness firms (agribusiness equipment, agrochemicals, seed companies and downstream industries) offer digital technologies that contribute to collecting data on agricultural production (weather, practices, yield). Actors in the agricultural sector, particularly those involved in conventional agriculture, then raise concerns about the risk of appropriating the value created by digital technologies, from which they themselves hope to profit. However, intermediary agricultural organizations retain a central role and are not excluded from this digitalization process, which they participate in building, guiding, evaluating and deciding on, for example, with collective investments, experiments or the more endogenous development of digital technologies and following other economic models.

Furthermore, the use of digital technologies for the capitalization, access, and exchange of knowledge is becoming increasingly widespread in the agricultural sector and can impact networks and the circulation of knowledge. Some uses are a continuation of existing knowledge-sharing dynamics among peer groups and reinforce internal exchanges within cooperatives, farmer groups, or other organizations that develop tools for knowledge exchange and capitalization. This confirms the importance of existing online communities for exchanges between farmers (Skaalsveen et al. 2020). Other uses create new exchange communities around topics of common interest, such as conservation agriculture, which sometimes extend beyond the traditional organizational or territorial boundaries of this type of exchange and operate in more horizontal dynamics, thus also confirming existing research (Compagnone and Pribetich 2017). These new exchange communities would compensate for the absence of existing communities through mechanisms that strengthen the sense of belonging and trust within these groups (Phillips et al. 2018). Information exchanges appear to complement face-to-face exchanges, which remain essential, particularly regarding ecologization practices (Burton and Riley 2018; Klerkx and Rose 2020). Overall, the consultation of technical websites is becoming more widespread, contributing to a diversification of information sources, which confirms the literature on the subject (Fielke et al. 2020). Our results show that knowledge exchange on the internet is particularly active for agricultural machinery (equipment reviews, repairs) and regulations, areas where knowledge is less specific to the local context. However, Internet usage does not currently appear to be disrupting information and knowledge networks, but it is, however, changing the geography of exchanges and involving new intermediaries and knowledge providers.

The development of digital technology thus appears as a factor reinforcing trends of expansion and specialization, at least in the large-scale crop farming sector, transforming work and also creating new sources of differentiation between farms and organizations within the agricultural sector (Van der Ploeg 2018). At the agricultural sector level, digitalization raises questions about the organization of businesses, the role of the state and the dynamics of knowledge circulation, without, for the moment, calling into question the role of intermediary organizations in the sector. The digital transformations of agricultural production models and the sector as a whole, by modifying the structural organization of farms, practices, work, knowledge processes and the role of the state and industries, may come to impact the ecologization of agriculture and the coexistence of different models.

7.3. The complex links between digital technology and environmental issues

Digital development is sometimes presented as a solution to environmental challenges, and sometimes as incompatible with them. But empirical data are lacking in order to inform this debate. This section revisits this controversy, showing how our analyses highlight oppositions between digitalization and ecologization, while also revealing possible hybridizations or articulations.

7.3.1. *The oppositions between digital technology and forms of ecological agriculture*

First, the results on the uses and perceptions of digital technology by stakeholders confirm the opposition between current uses of digital technology and the development of agroecology. Indeed, some technologies appear to be designed primarily according to the needs, techno-economic frameworks, knowledge and the decision-making processes of conventional production methods.

7.3.1.1. *Points of contention between digital technology and ecological agriculture*

7.3.1.1.1. Techniques and practices

Our results reveal forms of technical opposition, even technological incompatibilities, between digital technologies and strong ecologization practices. Several technologies only work with one or a few crops, which can encourage the increased production of that/those crop/crops rather than diversification, thus favoring monoculture-based systems. This confirms hypotheses put forward in the literature on this subject (Bronson 2018). Furthermore, the inputs considered in the design of these technologies are essentially conventional ones (seeds, pesticides, fertilizers), which can lead to malfunctions if they are used with alternative inputs (farmers' seeds, organic fertilizers, biocontrol), as reported by some farmers.

7.3.1.1.2. Objectives

Forms of conflict also emerge in terms of objectives. The objectives integrated into these technologies, and therefore into recommendations, generally remain based on maximizing yields and optimizing inputs (Oui 2021b). This can limit the capacity of these technologies to meet the diverse objectives of farmers, particularly those guiding alternative models (Bronson 2019). For example, some farmers are implementing forms of ecologization where maximizing input efficiency is not the primary objective, but where traceability and communication are paramount (e.g. quality wines). The optimization logic of certain technologies is reasoned in terms of risk management and securing production through pesticides, which may come into conflict with environmental objectives, as evidenced by work on the mechanisms of the Eco-phyto plan (Guichard et al. 2017). Furthermore, the optimization logic sometimes leads to rebound effects, with some farmers reporting increased input use when using certain DTPs, thus confirming hypotheses from the literature (Moschitz and Stolze 2018). In viticulture, our results show less incompatibility, which can be explained by the greater similarity in pesticide optimization issues between organic and conventional production methods (Serrano et al. 2019).

7.3.1.1.3. Reasoning

The coherence between digital technologies and ecologization practices does not appear solely as a problem of objectives, but also refers to a global line of reasoning. The results show, for example, that the basic unit of certain technologies is the technical itinerary of cultivation, which corresponds to a conventional way of thinking, but does not necessarily incorporate other ways of thinking, based on rotations, cover crop management or intercropping, interactions with landscape infrastructure, the territory or the agri-food system. Oppositions therefore exist in terms of technical reasoning, but also in terms of economic reasoning. Indeed, our interviews reveal ecologization practices that are integrated into forms of economic reasoning, which are different from the conventional model. Thus, farmers and agricultural organizations prioritize reducing investments and production costs, which can be incompatible with investing in high-tech digital equipment. Opposition can also be linked to the realm of ideas and values. For example, some aim to limit their farm's dependence on external factors or to strengthen local circularities (territorial bioeconomy, the reintroduction of livestock farming, market stabilization within territorial food systems, etc.). Others, finally, consider the relationship with natural objects as paramount, associated with a sensory connection that is not taken into account in the digital technologies deployed today, except in experiments that are still nascent (Javelle and Garcia 2022).

7.3.1.1.4. Organization and social model

Furthermore, opposition may be linked to the social model or the organization of forms of ecologization. For example, the individualization and commodification of advice, as well as the influence of practices associated with digital technologies, may be incompatible with the social aspect or the collective risk management that are promoted in agroecology (Stone 2022). Similarly, the increase in farm size, facilitated by digital technologies, may appear to contradict the social objectives of certain conceptions of agroecology such as the resettlement of farmers and the increase in the social fabric of rural areas (Lebrun et al. 2020).

7.3.1.2. *Digital technology and lock-in*

These opposing mechanisms in terms of technique, objective or reasoning mean that digital technologies do not support all forms of ecologization in the same way. Our results confirm that some digital technologies primarily meet the needs of weak ecologization based on input optimization. This optimization is necessary in all forms of ecologization. However, some forms of ecologization opt for practices that reduce inputs before even attempting to optimize them. This limits their appeal and their return on investment for input dose optimization technologies. Some technologies also respond to forms of symbolic ecologization, that is, based on justification and display rather than on changes in practices (Wolf and Wood 1997). Thus, the use of digital tools for regulatory traceability is becoming widespread: to track input applications and justify them. This justification makes it possible to circumvent general regulatory limits by justifying localized needs (fertilizer dose or spraying conditions, for example), as discussed in the literature (Oui 2021b). Our investigations show that these tools can even be used to circumvent regulations by modifying data to meet regulatory requirements, for example, by changing the date or dose of treatment. They are also used to perform the traceability required by labels or specifications, for example, for High Environmental Value (HVE) certification in viticulture or quality supply chains in the cereal industry. In our interviews, actors in the conventional model thus see digital technology as a way to "improve the image of agriculture". These technologies thus respond to an environmental legitimation, promoted in particular by downstream industries and public policies, and specific to the contemporary agri-food system (Wolf and Wood 1997; Prause et al. 2020). Environmental regulations and quantifications then appear as central drivers of the development of digital technology on farms (Koutsos and Menexes 2017; Oui 2021b).

Our results, derived from a cross-sectional analysis, provide information on the phenomena observed at a given point in time. However, they also offer avenues for analyzing longer term effects. If these technologies require an investment from the farmer using them, this can create lock-in effects. Kirat (1991) showed that the development of agricultural machinery led to increased complexity and interdependence between production components, creating irreversibilities and reinforcing the dominant paradigm. Our results suggest that certain digital technologies continue to increase complexity and interdependence, further locking in the dominant paradigm and its various components, including the use of conventional inputs. This lock-in phenomenon can be even more significant when these digital technologies are associated with industry constraints and standards (Magrini et al. 2016), or with advisory and regulatory structures that are not geared toward the strong ecologization of practices (Vanloqueren and Baret 2008; Labarthe 2010). Optimization technologies of the conventional model appear more and more as a way of defending and pursuing this model (Wolf and Buttel 1996).

Lock-ins can also be linked to the transformations in sector knowledge brought about by digitalization. The orientation of resources and knowledge systems toward digital development can sideline the development of knowledge on alternative ecologization strategies and lock in the technological trajectory (Cowan and Gunby 1996; Vanloqueren and Baret 2009). This lock-in can be exacerbated by the growing role of agribusiness firms in directing innovation. The privatization of consulting and the commodification of data and knowledge tend to lead to new forms of industrialization rather than ecologization (Wolf and Wood 1997). Furthermore, the use of digital technologies can be associated with forms of standardization and can limit experimentation and the search for new techniques (Lioutas et al. 2019), particularly when linked to forms of integration where the technical content is predefined by downstream actors. These forms of standardization can freeze technical pathways, which can limit forms of ecologization that involve their redesign.

7.3.1.3. *Digital technology and power*

This question of the oppositions, or even the incompatibility, between these trajectories can also be discussed from a more political, social and economic perspective.

First, the development of digital technology in the agricultural sector impacts power relations. The literature highlights the fact that data collection and analysis via digital technologies tends to strengthen the power of upstream and downstream industries (Wolf and Buttel 1996; Regan 2019; Barrett and Rose 2020; Prause et al. 2020). This leads to the appropriation of data and value by firms, the effects of which on ecologization are addressed rather briefly in the literature. Some studies emphasize the problems of firm dependence and of the lock-in of dominant models (Lioutas and Charatsari 2020), excluding other visions of sustainability and controlling the direction of technological innovation (Clapp and Ruder 2020). Some studies also raise the issue of the ability to repair, adapt and modify agricultural equipment according to individual objectives rather than those set by manufacturers and the industry (Carolan 2017b). Our field research highlights that the development of digital technology is partly a response to the industry's need for standardization and the control of supply chains, which can lead to rigid practices. The tools provided by upstream industries also appear to create new forms of dependence and lock-in on these firms and the agricultural model that they promote.

These power dynamics also influence the representations and principles that shape desirable futures and political decisions. Agricultural digital companies operate within the framework of techno-scientific promises (Joly 2015) by proposing digital agriculture as a solution to integrate productivist, ecological, and social challenges and offering a diversified, collaborative, and productive agriculture as a horizon, but also a competitive and sovereign AgTech sector (Martin and Schnebelin 2024). This promise is taken up and relayed by the press, in public policy documents, as well as by international organizations (Lajoie-O'Malley et al. 2020) and could tend toward a form of cultural hegemony that places digital technology at the heart of the transformations of society and the agricultural sector. These representations contribute to legitimizing solutions, building and evaluating public policies, and allocating resources by directing budgets. On the one hand, innovation policies may tend to fund digital and high-tech technologies at the expense of other, low-tech, organizational, social and institutional innovations that could directly promote agroecology (Barrett and Rose 2020). On the other hand, our work shows that power relations direct resources toward digital technology that responds to dominant forms of agriculture and their weak or symbolic ecologization, and thus reinforces existing power relations within the sector.

Digital technology thus appears to exacerbate certain divergences between agricultural models, whose organizations and individuals do not have the same expectations, do not perceive the same risks and do not use digital technology in the same way. This raises questions about the potential long-term exclusion of alternative agricultural models, which, for example, might no longer meet regulatory or buyer requirements, and which are based on digital technology.

7.3.2. *Potential hybridizations to be built*

Empirical analysis, and in particular the analysis of uses at the microeconomic level, however, tempers these effects of divergence and reinforcement. Our results indeed reveal forms of hybridization or convergence between the development of digital technology and a more or less pronounced ecologization of the agricultural sector.

An original and perhaps unexpected finding from fieldwork with large-scale crop farmers is the overrepresentation of mixed (organic-conventional) farms among those heavily reliant on digital technology. Digital technology appears to be a tool that facilitates the implementation of organic production methods while maintaining an organization, which is similar to conventional agriculture. Digital technology thus seems to promote forms of industrial ecologization and the conventionalization of the agroecological transition process, even drawing on references such as organic farming (Gkisakis and Damianakis 2020). It would participate in new forms of differentiation within organic agriculture in particular (Dufeu et al. 2020).

Some surveys reveal original or improvised uses of digital technologies and find ways to adapt these technologies to needs related to their ecological practices. These needs are linked to mechanical weeding, spatial and intra-plot crop diversification, and working with cover crops. The appropriation and instrumentation of the technical artifact allow for a better integration of its use with the knowledge and know-how of certain users, and improve its use in terms of sustainability (Moreiro 2017).

The arrival of "new entrants" into the sector, bringing different ideas and skills, raises specific challenges in this regard. These new entrants are characterized by higher levels of education than the average farmer, skills acquired through previous professional experience and a strong commitment to alternative forms of agriculture, particularly organic farming (Sutherland 2015). On the one hand, these new entrants have significant knowledge needs and less access to traditional advisory networks, making the internet all the more central to their access to knowledge (Sutherland et al. 2017). On the other hand, their skills facilitate the use, and even the development, of digital technologies. We observed, for example, a group of farmers developing a digital service thanks to a farmer who was formerly an IT specialist. We can also mention Atelier Paysan, a cooperative of collective interest focused on self-construction to promote peasant agroecology, which offers programming training to use the open-source electronics technology, such as Arduino (Giotitsas 2019). The spaces of friction thus created between the agricultural and digital worlds could give rise to an ecological and digital culture in which forms of hybridization can emerge.

The information and knowledge exchanged through digital technologies also offer opportunities for hybridization. The results, at different scales of analysis, show that the use of information and communication technologies can be associated with the development of ecologization, even to a pronounced degree. Indeed, these technologies support exchanges within organizations and between farmers. They are used to disseminate technical and scientific information and allow for the diversification and complementing of farmers' traditional sources of technical information. However, our results show that they do not systematically lead to changes in ecological practices and remain dependent on other forms of knowledge exchange. On the one hand, the use of digital technology for ecological knowledge has its limitations, since such knowledge also has tacit and localized forms, which involve local exchanges and concrete experiences (Burton and Riley 2018). On the other hand, these technologies contribute to a proliferation of information sources, the relevance and validity of which are difficult to verify. The interviews also highlighted the fact that information tends to simplify or idealize realities and fails to show failures. All of this limits its effective use by farmers. Digital technologies could also facilitate "On Farm Experimentation", or collaborative research projects, by offering new ways to exchange and capitalize on knowledge, as well as by creating analytical and evaluation tools that integrate ecological and landscape components (Lacoste et al. 2021).

A fourth aspect concerns new forms of autonomy which could facilitate agroecological transition. Some farmers use digital technology to source goods or services for agroecological practices, while others are developing new marketing channels. Digital technology thus offers new tools that could allow farmers to break free from certain dependencies and create new opportunities. For example, the use of digital technology can facilitate access to agricultural equipment for experimenting with agroecological practices (seed drills for no-till farming, cultivators, sorters, etc.) or access to local inputs (local seeds, organic fertilizers).

Digital technology also appears to be fostering convergence between models, particularly at the farm level. Common expectations are emerging, such as improved working conditions, autonomy and the ability to repair equipment. Information-sharing networks also seem to be transcending certain divisions between models, as do groups of farmers who share information about their profession on YouTube (Rénier et al. 2022). Common challenges are emerging around the image of agriculture and communication with consumers and local residents. However, as in other sectors or social spaces, there are effects of cognitive community enclosure, for example, around media outlets or social media communities, which explicitly reflect a productivist and anti-agribashing logic, and which only affect their immediate community. On the contrary, some actors, local authorities or media, are seeking a broader positioning at the interface of different models and are using digital technology to promote principles of coexistence (the BoCal platform supported by the Montpellier metropolitan area is an example).

The results from fieldwork suggest that what seems important for fostering digital technology which is compatible with the agroecological transition is both a systemic and multilevel approach. A systemic approach allows to take into account technical, knowledge-based and economic dimensions. A multi-level approach supports the integration of action at both individual and collective scales. The collective aspect can be mobilized at several stages. On the one hand, the collective facilitates learning and the appropriation of technologies, and therefore their more sustainable use (Badillo and Pélissier 2015; Moreiro 2017). Work on cooperative wineries shows that collective organizations are relevant intermediaries for addressing the challenges of technology mastery and appropriation by farmers. On the other hand, collective organization allows us to formulate needs and participate in the design of technologies to better integrate the diversity of needs (Bronson 2019).

7.4. Guiding innovation

This research highlights the importance of the direction of innovation. Indeed, innovation policies cannot be neutral: they are embedded in technological trajectories, respond to specific needs or interests and favor certain actors. The concept of directionality evokes this orientation of the innovation trajectory in a defined direction (Pigford et al. 2018; Hekkert et al. 2020). It highlights the fact that the innovation system is not merely a catalyst for all innovation, but also plays a role in guiding innovations along a trajectory, toward explicit or implicit objectives. Mechanisms for selecting innovations are at work within this system based on values and principles, economic forces, and institutional and social factors, whose interactions shape innovation trajectories (Dosi 1982; Malerba 2002). Changes will depend on the paradigm of the actors in the innovation system, who participate in guiding innovation according to knowledge, standards, existing innovations, etc. (Dockès 1990; Kirat 1991). It is therefore essential to consider this non-neutrality and to reflect on the factors that guide innovation and how to influence this direction.

Several recent research studies incorporate such reflections on the orientation of innovation systems to meet societal objectives. Some authors propose fostering niches of ecological innovation (Pigford et al. 2018). Others develop the concept of a "mission-oriented innovation system", which involves establishing innovation governance, defining a direction and coordinating technological, institutional, social and other types of innovation (Hekkert et al. 2020; Klerkx and Begemann 2020). However, this raises the problem of defining orientation: How can it be defined in the face of global challenges and an uncertain context, while remaining democratic?

Klerkx and Rose (2020) propose including a diversity of transition pathways in innovation policies. Others suggest a more participatory development of digital technologies to promote the agroecological transition (Wittman et al. 2020). But the direction of innovation also depends on power dynamics. Participatory or diversity-inclusive approaches do not eliminate power relations and strategic behaviors (Shove and Walker 2007). Actors with sufficient power can influence institutional change to facilitate innovations that serve their interests (Kukk et al. 2016).

This book highlights the fact that shifting digitalization toward ecologization requires taking into account intra-sectoral heterogeneity, power dynamics, and the plurality of needs and futures. It implies integrating technical issues into economic and political ones. The orientation of digitalization toward ecologization necessitates considering the potential complementarities between digital technologies and agroecology (Gkisakis and Damianakis 2020), but also offsetting power relations within innovation systems to promote a diversity of possibilities (Tsing 2017; Bulah et al. 2024).

Conclusion

The arrival of digital technology in agriculture is transforming the sector in its institutions, organizations, actors, knowledge and practices. Chapter 1, through a review based on scientific and grey literature, provides an initial description of the various dimensions of the digitalization of agriculture and the controversies it generates. These controversies raise the central question that guides this work: Does the development of digital technology in the agricultural sector promote or hinder its ecological transition? Chapter 2 draws on a review of more theoretical literature in the social sciences and economics to demonstrate the importance of a political economy perspective on innovation in order to conduct a systemic analysis integrating the different dimensions and scales of the sector's ecologization, as well as the heterogeneity of its trajectories. The conceptual framework of the innovation system, by integrating the notion of paradigm and by combining different scales of analysis, from the sector to individual practices, allows for such an analysis of agricultural transformations. Chapter 3 then proposes a way to implement this conceptual framework in a multi-scale analysis of the digitalization and ecologization of agriculture.

Chapter 4 describes the transformations of the agricultural innovation system through the development of digital technology and through the perceptions of actors who are linked to different paradigms of ecologization. Some perceptions are widely shared between the actors of different ecologization paradigms, such as the role of agricultural organizations in the support and implementation of digitalization, the data appropriation risks by digital companies, or the loss of autonomy, expectations for better working conditions and access to new knowledge. But differences divide the agricultural sector, far from the unifying vision generally held by digital companies. While organizations that are related to organic farming hope for digital support for observing, analyzing and systematically designing production systems, facilitating knowledge sharing or strengthening direct exchanges with consumers, those related to conventional agriculture hope that digital technology will facilitate traceability, will be a source of value for agricultural organizations, will limit negative impacts on the environment and will improve the image of agriculture. While the former fear standardization, a loss of know-how, of control and power, the latter identify risks of value appropriation and risks of limited use because of farmers lack of skills. Finally, organizations that adhere to organic or conventional paradigms are distinguished by the directions of digitalization that they envision. The former envisage a form of digitalization that, on one hand, strengthens exchanges and collaborations, and, on the other hand, supports the implementation of complex production systems based on ecosystem services. The latter envisage technological development aimed, on one hand, at reinforcing traceability processes and enhancing the value of production, and, on the other hand, at developing technologies to optimize and simplify production, and improve profitability, while limiting environmental impacts.

Chapter 5 highlights the microeconomic and material dimensions of digitalization. At the level of large-scale crop farms and their use of digital technologies, differences are observed between organic and conventional farms. Overall, use profiles differ according to the type of production model. Farms with a smaller economic size tend to be more diversified, practice organic farming or direct sales and are overrepresented among those that make limited use of digital technologies for production (DTP). Farms with more salaried labor and operating conventionally are overrepresented among those with medium DTP use. Larger farms, which provide agricultural services, have more labor or operate mixed systems (organic-conventional), are overrepresented among those making the most use of DTP. Mixed farms, producing both organic and conventional crops, show that digital technologies can facilitate forms of industrial ecological agriculture. These results also highlight that, beyond the production model (organic or conventional), it is primarily the degree of industrialization of the farm that characterizes its use of DTP. Digital technologies appear to be both facilitated by – and facilitators of – dynamics that characterize agricultural industrialization. These dynamics include specialization, concentration, wage labor, increased capital and the integration of farms with upstream and downstream industries. The results also reveal another dimension: subcontracting. The use of digital technologies accompanies the implementation of services. Conversely, the use of service providers can introduce digital technologies into farms, but this does not appear to be a decisive factor in choosing services' providers. The interrelations between digital technology use, service provision and the ecologization of agriculture remain to be further explored.

The development of digital technologies in agriculture also reflects broader sectoral dynamics. The use of digital tools for technical information is becoming widespread among farmers. Differences in use are primarily linked to individual and network characteristics. Farmers advised by cooperatives or seed companies use digital information and communication technologies (ICTs) less, whereas those who are part of a CETA or other farmer exchange groups rely more on getting informed digitally. Younger farmers or those who are members of agricultural unions use social networks more frequently to access information.

Chapter 6 focuses on a specific type of organization in the agricultural sector: cooperatives, seen both as intermediary organizations within the agricultural innovation system and as systems of interaction between producers, to study forms of coordination and links between sectoral changes and individual practices. In viticulture, digital technology use in the Occitanie region appears primarily as a collective dynamic, driven by downstream requirements, upstream technological offerings and internal cooperative initiatives to increase knowledge-sharing, communication with consumers and even governance dynamics. Digitalization is mainly implemented to support traceability, whether regulatory or linked to specific standards, reflecting a weak or even symbolic ecologization. It is also widely used for communication in a sector where quality and value creation are largely built on image and territorial and producer identity. The example of cooperative wineries more broadly suggests that agricultural cooperatives can become spaces for shaping digitalization and ecological trajectories, developing individual and collective capacities, and articulating these trajectories.

Digital agriculture appears as an "umbrella" term, covering a wide range of technologies. It is described as a revolution, a "new agricultural model" and a transformation of the sector beyond the divides between large and small farms, between organic and conventional farms, and between those producing for local markets and those producing for global markets. Yet, this book shows that the development of digital technologies in agriculture generates a diversity of perceptions, uses and trajectories, which appear to reinforce distinctions between paradigms and accompany mechanisms of differentiation among agricultural production models, continuing the sectoral transformations that have occurred since World War II. This book thus presents three main empirical contributions:

– The digitalization of agriculture brings together diverse trajectories and is not linear, with phenomena of abandonment, learning, adaptation, reinforcement, etc. The use of digital technology and the digitalization of the sector more generally does not appear to be the result of so-called "pioneering" behaviors, but depends on the diversity of paradigms and ecologization trajectories.

– Digitalization accompanies the industrialization of agriculture, characterized by productive specialization; increasing reliance on wage labor; intensification; farm enlargement; growing use of external products, energy and services; and integration into value chains. However, digitalization is also accompanied by more disruptive transformations regarding knowledge circulation dynamics in the sector, and it questions the roles of the state, firms and collective or intermediary organizations.

– Current digitalization shows several forms of opposition to strong ecologization, whether in technical terms, objectives, reasoning, temporal dynamics, or political and social issues. However, hybridizations of digitalization with forms of industrial ecological agriculture appear possible, as well as a more global transformation of digitalization through rethinking its technical, economic and political models.

This book emphasizes that the development of digital technologies in agriculture is not neutral. It calls for reflection on modes of innovation governance, the current dominant form of which is the regime of technoscientific promises. This regime is based on problematizing issues and presenting technologies as the only way to address them, proposing radical novelty while building credibility, relying on the naturalization of the proposed technological evolution and the support of a circle of expertise. This mode of governance creates a state of necessity, divides society between technology promoters and users, excludes alternative solutions, and sidelines reflection on the societal issues of innovations. It therefore seems necessary to renew this mode of governance. Innovation policies could be based on studies of the effects of these technological developments and critical reflection that integrates real economic and social issues, rather than mere promises. Furthermore, research and innovation policies could encourage forms of collective experimentation and facilitate the emergence of alternative solutions in order to avoid technological lock-ins.

The analytical framework developed in this book and its associated method appear relevant for analyzing innovation more generally across diverse contexts. This framework proposes crossing levels of analysis to study innovations, which both impact and are impacted by sectoral organizations, institutions and individual practices. On the one hand, this framework allows for examining the intertwinings and co-evolutions of innovation systems and individual practices that occur during innovation trajectories. It shows how institutional developments are appropriated by individual practices, and how the latter can gradually influence the institutional framework (Laurent et al. 2021). This method reveals the concrete effects associated with an innovation, going beyond promises, by confronting actual uses and practices with those anticipated or perceived by actors. It can reveal reconfigurations, rearrangements, differentiated uses, rebound effects or combinations/interactions that may even contradict expected outcomes (Rose et al. 2018; Visser et al. 2021). The combination of analysis levels proposed here is relevant for highlighting historical evolutions and trajectories, emphasizing mechanisms of reinforcement, exclusion, lock-in and so on. Crossing levels of analysis to grasp the complexity and interrelatedness of factors influencing innovation processes allows for understanding temporal dynamics, which can then be explored in more detail through longitudinal analyses. These multi-level innovation analyses thus help to understand the barriers, levers, heterogeneities, strengths and limits of transformations associated with innovations.

This book addresses how to support innovations in agriculture to account for the diversity of needs, contexts and practices required for the ecological and social transformation of the agricultural sector. It calls for considering sectoral heterogeneities, power relations and the plurality of needs and futures, as well as integrating technical questions into economic and political issues in order to counterbalance power relations within innovation systems and promote a polyphonic future.

References

ACI (1995). Alliance Coopérative Internationale [Online]. Available at: https://www.ica.coop/fr/coop%C3%A9ratives/identite-cooperative [Accessed 4 April 2022].

ADEME, Deloitte Développement Durable, Deloitte in Extenso, G-SCOP, SATIE (2017). Industrie du futur : comment allier transition numérique et transition énergétique et écologique ? Report [Online]. Available at: https://librairie.ademe.fr/produire-autrement/1766-comment-allier-transition-numerique-et-transition-energetique-et-ecologique-.html [Accessed 23 September 2021].

Agence Bio (2020). Les chiffres clés [Online]. Available at: https://www.agencebio.org/vos-outils/les-chiffres-cles/ [Accessed 6 September 2021].

AgFunder (2021). 2021 AgFunder AgriFoodTech Investment Report.

Aghion, P. and Antonin, C. (2017). Progrès technique et croissance depuis la crise. *Revue de l'OFCE*, 153(4), 63–78.

Agriculture (2021). Grand plan d'investissement : la gamme de financements du volet agricole [Online]. Available at: https://agriculture.gouv.fr/gpi [Accessed 22 September 2021].

agriculture.gouv (2016). Infographie – Les chiffres clés de l'agriculture connectée [Online]. Available at: https://agriculture.gouv.fr/infographie-les-chiffres-cles-de-lagriculture-connectee [Accessed 28 July 2021].

AgroTIC Observatory (2017). Observatoire des usages de l'agriculture numérique. Une action soutenue par la Chaire AgroTIC et l'Institut Convergences #DigitAg [Online]. Available at: https://agrotic.org/observatoire/ [Accessed 14 February 2020].

Akrich, M. (1989). La construction d'un système sociotechnique. Esquisse pour une anthropologie des techniques. *Anthropologie et sociétés*, 13(2), 31–54.

Allaire, G. (1988). Le modèle de développement agricole des années 1960. *Économie rurale*, 184(1), 171–181.

Allaire, G. (2002). L'économie de la qualité, en ses secteurs, ses territoires et ses mythes. *Géographie Économie Société*, 4(2), 155–180.

Allaire, G. and Boyer, R. (1995). *La grande transformation de l'agriculture : lectures conventionnalistes et régulationnistes*. Quæ, Versailles.

Amable, B. (2000). Institutional complementarity and diversity of social systems of innovation and production. *Review of International Political Economy*, 7(4), 645–687.

Amable, B. (2005). Les cinq capitalismes. Diversité des systèmes économiques et sociaux dans la mondialisation. Report, Paris [Online]. Available at: https://ideas.repec.org/p/hal/cesptp/halshs-00310487.html [Accessed 18 November 2021].

Ansaloni, M. and Fouilleux, E. (2006). Changement de pratiques agricoles. Acteurs et modalités d'hybridation technique des exploitations laitières bretonnes. *Économie rurale. Agricultures, alimentations, territoires*, 292, 3–17.

Ansaloni, M. and Smith, A. (2021). *L'expropriation de l'agriculture française. Pouvoirs et politiques dans le capitalisme contemporain*. Éditions du Croquant, Vulaines-sur-Seine.

Arthur, W.B. (1989). Competing technologies, increasing returns, and lock-in by historical events. *The Economic Journal*, 99(394), 116–131.

Aspexit (2021). Agriculture & numérique : prenons-nous vraiment la bonne direction ? [Online]. Available at: https://www.aspexit.com/agriculture-numerique-prenons-nous-vraiment-la-bonne-direction/ [Accessed 29 March 2022].

Atelier Paysan (2021). *Reprendre la terre aux machines. Manifeste pour une autonomie paysanne et alimentaire*. Le Seuil, Paris.

Atkinson, R. and Flint, J. (2001). Accessing hidden and hard-to-reach populations: Snowball research strategies. *Social Research Update*, 33(1), 1–4.

Audretsch, D., Colombelli, A., Grilli, L., Minola, T., Rasmussen, E. (2020). Innovative startups and policy initiatives. *Research Policy*, 49(10), 104027.

Aulagnier, A. and Goulet, F. (2017). Des technologies controversées et de leurs alternatives. Le cas des pesticides agricoles en France. *Sociologie du travail*, 59(3) [Online]. Available at: http://journals.openedition.
org/sdt/840 [Accessed 27 February 2020].

Aune, J.B. (2012). Conventional, organic and conservation agriculture: Production and environmental impact. In *Agroecology and Strategies for Climate Change*, Lichtfouse, E. (ed.). Springer, Dordrecht. doi: 10.1007/978-94-007-1905-7_7.

Avril, P., Ridier, A., Samson, É. (2021). PAC et environnement : les exploitations les plus vertueuses sont-elles récompensées ? *Économie rurale. Agricultures, alimentations, territoires* [Online]. Available at: https://journals.openedition.org/economierurale/9185 [Accessed 30 April 2022].

Ayerdi Gotor, A., Marraccini, E., Leclercq, C., Scheurer, O. (2020). Precision farming uses typology in arable crop-oriented farms in northern France. *Precision Agriculture*, 21(1), 131–146.

Ayre, M., Mc Collum, V., Waters, W., Samson, P., Curro, A., Nettle, R., Paschen, J.-A., King, B., Reichelt, N. (2019). Supporting and practising digital innovation with advisers in smart farming. *NJAS: Wageningen Journal of Life Sciences*, 90(91), 100302.

Babinet, G. (2021). Comment concilier transition écologique et numérique? Institut Montaigne [Online]. Available at: https://www.institutmontaigne.org/blog/comment-concilier-transition-ecologique-et-numerique [Accessed 23 September 2021].

Badillo, P.-Y. and Pélissier, N. (2015). Usages et usagers de l'information numérique. Renouvellement des problématiques et nouveaux enjeux pour les SIC. *Revue française des sciences de l'information et de la communication* [Online]. Available at: http://journals.openedition.org/rfsic/1448 [Accessed 31 March 2021].

Bages, R. (1992). Pratiques de l'informatique chez les agriculteurs et degrés d'appropriation. Une enquête dans la région Midi-Pyrénées. *Économie rurale*, 210(1), 10–16.

Bai, C., Dallasega, P., Orzes, G., Sarkis, J. (2020). Industry 4.0 technologies assessment: A sustainability perspective. *International Journal of Production Economics*, 229, 107776.

Balafoutis, A., Beck, B., Fountas, S., Vangeyte, J., Wal, T.V. der Soto, I., Gómez-Barbero, M., Barnes, A., Eory, V. (2017). Precision agriculture technologies positively contributing to GHG emissions mitigation, farm productivity and economics. *Sustainability*, 9(8), 1339.

Bareille, F., Bonnet-Beaugrand, F., Duvaleix-Tréguer, S. (2017). Objectives' alignment between members and agricultural cooperatives. *Review of Agricultural, Food and Environmental Studies*, 98(1), 75–91.

Barnes, A.P., Soto, I., Eory, V., Beck, B., Balafoutis, A., Sánchez, B., Vangeyte, J., Fountas, S., van der Wal, T., Gómez-Barbero, M. (2019). Exploring the adoption of precision agricultural technologies: A cross regional study of EU farmers. *Land Use Policy*, 80, 163–174.

Barrett, H., Rose, D.C. (2020). Perceptions of the fourth agricultural revolution: What's in, what's out, and what consequences are anticipated? *Sociologia Ruralis*, 62, 162–189.

Bartoli, P. and Boulet, D. (1989). Dynamique et régulation de la sphère agroalimentaire : l'exemple viticole. PhD Thesis, Université Montpellier 1, Montpellier [Online]. Available at: https://hal.inrae.fr/tel-02856028 [Accessed 7 March 2022].

Bechtet, N. (2022). L'influence de la digitalisation sur les institutions de contrôle de l'efficacité des technologies en agriculture. Le cas des outils numériques d'aide à la décision pour la fertilisation azotée. In *Agriculture et numérique. Digitalisation des systèmes agri-alimentaires. Perspectives SHS.* Société française d'economie rurale, Montpellier.

Bechtet, N. (2023). How do advisory suppliers support farmers in evaluating a digital innovation? A case study on decision support tools for fertilizer application in France. *Journal of Innovation Economics & Management*, 42(3), 73–101.

Bellon, S. and Penvern, S. (2014). Organic food and farming as a prototype for sustainable agricultures. In *Organic Farming, Prototype for Sustainable Agricultures*, Bellon, S. and Penvern, S. (eds). Springer, Dordrecht. doi: 10.1007/978-94-007-7927-3_1.

Bellon, S., Lamine, C., Ollivier, G., de Abreu, L.S. (2011). The relationships between organic farming and agroecology [Online]. Available at: http://www.alice.cnptia.embrapa.br/handle/doc/917959 [Accessed 12 February 2020].

Bellon Maurel, V. and Huyghe, C. (2017). Putting agricultural equipment and digital technologies at the cutting edge of agroecology. *OCL*, 24(3), D307.

Bellon Maurel, V., Garcia, F., Brossard, L., Mitton, N., Termier, A. (2022). Agriculture et numérique. Tirer le meilleur du numérique pour contribuer à la transition vers des agricultures et des systèmes alimentaires durables. Report, INRAE.

Bellon-Maurel, V., Piot-Lepetit, I., Lachia, N., Tisseyre, B. (2023). Digital agriculture in Europe and in France: Which organisations can boost adoption levels? *Crop and Pasture Science*, 74(6), 573–585.

Bergek, A., Jacobsson, S., Carlsson, B., Lindmark, S., Rickne, A. (2008). Analyzing the functional dynamics of technological innovation systems: A scheme of analysis. *Research Policy*, 37, 407–429.

Berkhout, F. and Hertin, J. (2004). De-materialising and re-materialising: Digital technologies and the environment. *Futures*, 36(8), 903–920.

Berriet-Solliec, M. (1997). Institutions et territoires : vers le renouvellement des politiques publiques ? Les interventions des départements et des régions dans le secteur agricole. PhD Thesis, Université de Montpellier, Montpellier [Online]. Available at: http://www.theses.fr/1997ENSA0005 [Accessed 21 April 2022].

Bertini, M.-J. (2009). *Ni d'Eve ni d'Adam : défaire la différence des sexes*. Max Milo, Paris.

Beus, C.E. and Dunlap, R.E. (1990). Conventional versus alternative agriculture: The paradigmatic roots of the debate. *Rural Sociology*, 55(4), 590–616.

Birner, R., Daum, T., Pray, C. (2021). Who drives the digital revolution in agriculture? A review of supply-side trends, players and challenges. *Applied Economic Perspectives and Policy*, 43, 1260–1285.

Boinon, J.-P. (2011). Les politiques foncières agricoles en France depuis 1945. *Economie et Statistique/Economics and Statistics*, 444, 19–37.

Bonneuil, C. and Hochereau, F. (2008). Gouverner le "progrès génétique". Biopolitique et métrologie de la construction d'un standard variétal dans la France agricole d'après-guerre. *Annales. Histoire, Sciences Sociales*, 63(6), 1303–1340.

Bonny, S., (2017). High-tech agriculture or agroecology for tomorrow's agriculture? *Harvard College Review of Environment & Society*, 4, 28–34.

Bordes, J.-P. (2017). Numérique et agriculture de précision. *Annales des Mines. Responsabilite et environnement*, 87(3), 87–93.

Bourgeois, A. and Sébillotte, M. (1978). Réflexion sur l'évolution contemporaine des exploitations agricoles. *Économie rurale*, 126(1), 17–28.

Bournigal, J.-M. (2016). Big Data : "Oui, nous changeons de modèle agricole !" Ministère de l'agriculture et de l'alimentation [Online]. Available at: https://agriculture.gouv.fr/big-data-oui-nous-changeons-de-modele-agricole [Accessed 20 September 2021].

Bournigal, J.-M., Houllier, F., Lecouvey, P., Pringuet, P. (2015). Agriculture Innovation 2025 : 30 projets pour une agriculture compétitive & respectueuse de l'environnement. Report [Online]. Available at: https://agriculture.gouv.fr/sites/minagri/files/rapport-agriculture-innovation2025.pdf.

Boyer, R. (1989). Histoire des techniques et théories economiques. *Cepremap*, 8908, 1–33.

Boyer, R. and Freyssenet, M. (2000). *Les modèles productifs*. La Découverte, Paris.

Boyer, J. and Touzard, J.-M. (2017). Research, innovation and competitiveness of French vineyards: An analysis by the institutions. *Systèmes alimentaires*, 1, 69–96.

Boyer, R., Chavance, B., Godard, O. (1994). *Les figures de l'irréversibilité en économie*. EHESS, Paris.

Brandtzaeg, P.B., Heim, J., Karahasanovic, A. (2011). Understanding the new digital divide–A typology of Internet users in Europe. *International Journal of Human-Computer Studies*, 69(3), 123–138.

Brechet, J.-P., Schieb-Bienfait, N. (2006). Projets et pouvoirs dans les régulations concurrentielles. La question de la morphogenèse d'une filière d'agriculture biologique. *Revue d'économie industrielle*, 113, 9–29. doi: 10.4000/rei.233.

Briggeman, B.C. and Whitacre, B.E. (2010). Farming and the internet: Reasons for non-use. *Agricultural and Resource Economics Review*, 39(3), 571–584.

Bronner, G. and Géhin, É. (2010). *L'inquiétant principe de précaution*. PUF, Paris.

Bronson, K. (2018). Smart farming: Including rights holders for responsible agricultural innovation. *Technology Innovation Management Review*, 8(2), 7–14.

Bronson, K. (2019). Looking through a responsible innovation lens at uneven engagements with digital farming. *NJAS: Wageningen Journal of Life Sciences*, 90–91, 100294.

Bronson, K. and Knezevic, I. (2016). Big Data in food and agriculture. *Big Data & Society*, 3(1), 2053951716648174.

Bui, S., Cardona, A., Lamine, C., Cerf, M. (2016). Sustainability transitions: Insights on processes of niche-regime interaction and regime reconfiguration in agri-food systems. *Journal of Rural Studies*, 48, 92–103.

Bulah, B.M., van Mierlo, B., Beumer, K., Gerritsen, A.L., Negro, S.O., Hekkert, M.P., Klerkx, L. (2024). Diversity and directionality: Friends or foes in sustainability transitions? *Science and Public Policy*, 51, 1075–1092.

Burton, R.J.F. and Riley, M. (2018). Traditional ecological knowledge from the internet? The case of hay meadows in Europe. *Land Use Policy*, 70, 334–346.

Busse, M., Doernberg, A., Siebert, R., Kuntosch, A., Schwerdtner, W., König, B., Bokelmann, W. (2014). Innovation mechanisms in German precision farming. *Precision Agriculture*, 15(4), 403–426.

Busse, M., Schwerdtner, W., Siebert, R., Doernberg, A., Kuntosch, A., König, B., Bokelmann, W. (2015). Analysis of animal monitoring technologies in Germany from an innovation system perspective. *Agricultural Systems*, 138, 55–65.

Buttel, F.H. (1994). Agricultural change, rural society, and the state in the late twentieth century: Some theoretical observations. In *Agricultural Restructuring and Rural Change in Europe*, Symes, D. and Jansen, A.J. (eds). Agricultural University, Wageningen.

Byé, P., Chanaron, J.-J., Perrin, J. (1989). Les déterminants de l'innovation en agriculture à travers la littérature sur le machinisme et les engrais. *Cahiers d'économie et sociologie rurales*, 10, 65–96.

Candemir, A., Duvaleix, S., Latruffe, L. (2021). Agricultural cooperatives and farm sustainability – A literature review. *Journal of Economic Surveys*, 35(4), 1118–1144.

Caplat, J. (2014). *Changeons d'agriculture : réussir la transition*. Actes Sud, Paris.

Carbonell, I.M. (2016). The ethics of big data in big agriculture. *Internet Policy Review*, 5(1) [Online]. Available at: https://policyreview.info/articles/analysis/ethics-big-data-big-agriculture [Accessed 16 September 2019].

Cardon, D. (2010). *La démocratie Internet : promesses et limites*. Le Seuil, Paris [Online]. Available at: http://id.erudit.org/iderudit/1008225ar [Accessed 21 February 2022].

Carolan, M. (2017a). Agro-digital governance and life itself: Food politics at the intersection of code and affect. *Sociologia Ruralis*, 57(S1), 816–835.

Carolan, M. (2017b). Publicising food: Big data, precision agriculture, and co-experimental techniques of addition. *Sociologia Ruralis*, 57(2), 135–154.

Carolan, M. (2018). "Smart" farming techniques as political ontology: Access, sovereignty and the performance of neoliberal and not-so-neoliberal worlds. *Sociologia Ruralis*, 58(4), 745–764.

Carolan, M. (2020). Automated agrifood futures: Robotics, labor and the distributive politics of digital agriculture. *The Journal of Peasant Studies*, 47(1), 184–207.

Carson, R. (1962). *Silent Spring*. Houghton Mifflin Harcourt, Boston.

Casilli, A.A. (2019). *En attendant les robots. Enquête sur le travail du clic*. Média Diffusion, Paris.

Ceruzzi, P.E. (2012). *Computing: A Concise History*. MIT Press, Cambridge.

Chambre d'agriculture Occitanie (2021). Agriscopie 2021 : Viticulture [Online]. Available at: https://occitanie.chambre-agriculture.fr/productions-techniques/viticulture/agriscopie-2020-viticulture/ [Accessed 5 March 2022].

Chapus, Q., Potier, V., Brailly, J. (2023). Créations de plateformes dans le secteur agricole et logiques relationnelles : du découplage au ré-encastrement ? [Online]. Available at: https://hal.science/hal-03991732 [Accessed 12 February 2024].

Chavance, B. (2012). L'économie institutionnelle. Report, Post-Print [Online]. Available at: https://ideas.repec.org/p/hal/journl/halshs-00945936.html [Accessed 2 March 2022].

Chiffoleau, Y. (2005). Learning about innovation through networks: The development of environment-friendly viticulture. *Technovation*, 25(10), 1193–1204.

Chiffoleau, Y. and Touzard, J.-M. (2014). Understanding local agri-food systems through advice network analysis. *Agriculture and Human Values*, 31(1), 19–32.

Chiffoleau, Y., Dreyfus, F., Touzard, J.-M. (2008). Ethics in French wine cooperatives: Part of a social movement? In *Creating Food Futures*, Farnworth, C.R. and Jiggins, J. (eds). Routledge, London.

Cholez, C., Magrini, M.-B., Galliano, D. (2017). Les contrats de production en grandes cultures. Coordination et incitations par les coopératives. *Économie rurale*, 360, 65–83.

Cholez, C., Magrini, M.-B., Galliano, D. (2020). Exploring inter-firm knowledge through contractual governance: A case study of production contracts for faba-bean procurement in France. *Journal of Rural Studies*, 73, 135–146.

Chomel, C. (2019). Loi Egalim : quel avenir pour les spécificités des coopératives agricoles ? *RECMA*, 354(4), 120–130.

Ciruela-Lorenzo, A.M., Del-Aguila-Obra, A.R., Padilla-Meléndez, A., Plaza-Angulo, J.J. (2020). Digitalization of agri-cooperatives in the smart agriculture context. Proposal of a digital diagnosis tool. *Sustainability*, 12(4), 1325.

Clapp, J. and Ruder, S.-L. (2020). Precision technologies for agriculture: Digital farming, gene-edited crops, and the politics of sustainability. *Global Environmental Politics*, 20(3), 49–69.

Cleaver, F. (2002). Reinventing institutions: Bricolage and the social embeddedness of natural resource management. *European Journal of Development Research*, 14(2), 11–30.

CoFarming (2017). Livre Blanc: #CoFarming : Le faire ensemble en réseau. Report, CoFarming, Paris.

Cohen, A.G. (2017). Des lois agronomiques à l'enquête agroécologique. Esquisse d'une épistémologie de la variation dans les agroécosystèmes. Tracés. *Revue de sciences humaines*, 33, 51–72.

Cohendet, P. and Llerena, P. (1999). La conception de la firme comme processeur de connaissances. *Revue d'économie industrielle*, 88(1), 211–235.

Cohendet, P., Grandadam, D., Simon, L., Capdevila, I. (2014). Epistemic communities, localization and the dynamics of knowledge creation. *Journal of Economic Geography*, 14(5), 929–954.

Compagnone, C. and Pribetich, J. (2017). Quand l'abandon du labour interroge les manières d'être agriculteur. *Revue française de socio-économie*, 18(1), 101–121.

Confédération Paysanne (2021). Une agriculture Robotique, Numérique et Génétique pour 2030, sans paysans ni paysannes ? La Confédération paysanne de la Drôme s'est mobilisée à Bourg-lès-Valence [Online]. Available at: https://www.confederationpaysanne.fr/actu.php?id=12057&PHPSESSID=akn85ob85sfktsv3p34u7l9ai2 [Accessed 10 February 2022].

Connor, D.J. (2008). Organic agriculture cannot feed the world. *Field Crops Research*, 106(2), 187–190.

Conti, C., Zanello, G., Hall, A. (2021). Why are agri-food systems resistant to new directions of change? A systematic review. *Global Food Security*, 31, 100576.

Cook, M.L. (2018). A life cycle explanation of cooperative longevity. *Sustainability*, 10(5), 1586.

Cooke, P., Gomez Uranga, M., Etxebarria, G. (1997). Regional innovation systems: Institutional and organisational dimensions. *Research Policy*, 26(4), 475–491.

Coriat, B. and Dosi, G. (2002). The institutional embeddedness of economic change: An appraisal of the "evolutionary" and "regulationist" research programmes. In *A Modern Reader in Institutional and Evolutionary Economics*, Hodgson, G.M. (ed.). Edward Elgar Publishing, London [Online]. Available at: http://www.elgaronline.com/view/9781840644746.00015.xml [Accessed 3 May 2022].

Cornilleau, L. and Joly, P.-B. (2014). La révolution verte, un instrument de gouvernement de la "faim dans le monde". Une histoire de la recherche agricole internationale. In *Le gouvernement des technosciences*, Pestre, D. (ed.). La Découverte, Paris, 171–201.

Coulomb, P. and Nallet, H. (1980). *Le syndicalisme agricole et la création du paysan modèle*. FeniXX, Paris.

Cowan, R. and Gunby, P. (1996). Sprayed to death: Path dependence, lock-in and pest control strategies. *The Economic Journal*, 106(436), 521–542.

Cusinato, A. (2007). Quelle approche pour l'économie informelle ? De l'individualisme méthodologique à la perspective institutionnaliste. *Commission suisse pour l'Unesco*, 1, 65–88.

Damave, M.-C. (2017). Tous acteurs de la transition numérique agricole. Report, AgriDées, Paris [Online]. Available at: https://www.agridees.com/publication/acteurs-de-transition-numerique-agricole/.

Daniels, A.K. (1987). Invisible work. *Social Problems*, 34, 403–415.

Darnhofer, I., Lindenthal, T., Bartel-Kratochvil, R., Zollitsch, W. (2010). Conventionalisation of organic farming practices: From structural criteria towards an assessment based on organic principles. A review. *Agronomy for Sustainable Development*, 30(1), 67–81.

Darré, J.-P. (1996). *L'invention des pratiques dans l'agriculture : vulgarisation et production locale de connaissance*. Karthala Editions, Paris.

Darré, J.-P., Le Guen, R., Lemery, B. (1989). Changement technique et structure professionnelle locale en agriculture. *Économie rurale*, 192(1), 115–122.

David, P.A. (1985). Clio and the economics of QWERTY. *The American Economic Review*, 75(2), 332–337.

Daviron, B. (2021). Aux origines de l'agriculture industrielle. In *Une écologie de l'alimentation*, Bricas, N.D. and Conaré Walser, M. (eds). Quæ, Versailles [Online]. Available at: https://www.quae-open.com/extract/646 [Accessed 18 April 2022].

Dehousse, F. and Vincent, P. (1998). L'eternelle réforme de la politique agricole commune et les limites d'agenda 2000. *Studia Diplomatica*, 51(5), 1–145.

Del Corso, J.-P., Kephaliacos, C., Plumecocq, G. (2015). Legitimizing farmers' new knowledge, learning and practices through communicative action: Application of an agro-environmental policy. *Ecological Economics*, 117, 86–96.

Deverre, C., de Sainte Marie, C. (2008). L'écologisation de la politique agricole européenne. Verdissement ou refondation des systèmes agro-alimentaires ? *Cahiers d'économie et sociologie rurales*, 89, 83–104.

DiMaggio, P., Hargittai, E. (2001). From the "digital divide" to "digital inequality": Studying internet use as penetration increases. *Center for Arts and Cultural Policy Studies*, 4, 1–25.

Diry, J.-P. (1988). Agriculture industrielle et agriculture industrialisée (Industrial agriculture and industrialized agriculture). *Bulletin de l'association de géographes français*, 65(2), 125–137.

Djellal, F. (1995). *Changement technique et conseil en technologie de l'information.* L'Harmattan, Paris.

Dobrov, G.M. (1979). La technologie en tant qu'organisation. *Revue internationale des sciences sociales*, XXXI(4), 628–649.

Dockès, P. (1990). Formation et transferts des paradigmes sociotechniques. *Revue française d'économie*, 5(4), 29–82.

Dosi, G. (1982). Technological paradigms and technological trajectories: A suggested interpretation of the determinants and directions of technical change. *Research Policy*, 11(3), 147–162.

Dosi, G., Freeman, C., Nelson, R., Silverberg, G., Soete, L. (1988). Technical change and economic theory. Report, LEM [Online]. Available at: https://econpapers.repec.org/bookchap/ssalembks/dosietal-1988.htm [Accessed 27 August 2020].

Dufeu, I., Le Velly, R., Bréchet, J., Loconto, A. (2020). Can standards save organic farming from conventionalisation? Dynamics of collective projects and rules in a French organic producers' organisation. *Sociologia Ruralis*, 60(3), 621–638.

Dufva, T. and Dufva, M. (2018). Grasping the future of the digital society. *Futures*, 107, 17–28 [Online]. Available at: https://linkinghub.elsevier.com/retrieve/pii/S0016328717302252 [Accessed 20 December 2018].

Dujarier, M.-A. (2014). *Le travail du consommateur. De McDo à eBay : comment nous coproduisons ce que nous achetons*. La Découverte, Paris.

Dumez, H. (2013). *Méthodologie de la recherche qualitative. Les questions clés de la démarche compréhensive*. Vuibert, Paris [Online]. Available at: https://books.google.fr/books?hl=fr&lr=&id=u8xHDAAAQBAJ&oi=fnd&pg=PT2&dq=M%C3%A9thodologie+de+la+recherche+qualitative+Dumez&ots=2WdzCqrFiX&sig=-ARqWeLfotwpyaJVt8r6YqXPIng#v=onepage&q=M%C3%A9thodologie%20de%20la%20recherche%20qualitative%20Dumez&f=false [Accessed 30 September 2019].

Durand, C. (2020). *Technoféodalisme : critique de l'économie numérique*. La Découverte, Paris.

Duru, M., Therond, O., Fares, M. (2015). Designing agroecological transitions. A review. *Agronomy for Sustainable Development*, 35(4), 1237–1257.

Duvaleix-Tréguer, S., Ledos, F., Lepetit, G. (2019). L'usage du numérique, facilitateur de la gouvernance ? *RECMA*, 351(1), 68–80.

Eastwood, C.R., Chapman, D.F., Paine, M.S. (2012). Networks of practice for co-construction of agricultural decision support systems: Case studies of precision dairy farms in Australia. *Agricultural Systems*, 108, 10–18.

Eastwood, C., Klerkx, L., Nettle, R. (2017). Dynamics and distribution of public and private research and extension roles for technological innovation and diffusion: Case studies of the implementation and adaptation of precision farming technologies. *Journal of Rural Studies*, 49, 1–12.

Eastwood, C., Ayre, M., Nettle, R., Dela Rue, B. (2019). Making sense in the cloud: Farm advisory services in a smart farming future. *NJAS: Wageningen Journal of Life Sciences*, 90–91, 100298.

Edquist, C. (2013). *Systems of Innovation: Technologies, Institutions and Organizations*. Routledge, London [Online]. Available at: https://www.taylorfrancis.com/books/9780203357620 [Accessed 17 June 2019].

Etikan, I. (2016). Comparison of convenience sampling and purposive sampling. *American Journal of Theoretical and Applied Statistics*, 5(1), 1.

Fairbairn, M. and Guthman, J. (2020). Agri-food tech discovers silver linings in the pandemic. *Agriculture and Human Values*, 37(3), 587–588.

FAO, ITU (2016). E-agriculture strategy guide. Report, FAO, ITU [Online]. Available at: http://www.fao.org/documents/card/en/c/24f624ea-7891-45e8-9b24-66cbf13f004d/ [Accessed 13 July 2021].

Faure, G., Chiffoleau, Y., Goulet, F., Temple, L., Touzard, J.M. (2018). *Innovation et développement dans les systèmes agricoles et alimentaires*. Quæ, Versailles.

Fielke, S.J., Garrard, R., Jakku, E., Fleming, A., Wiseman, L., Taylor, B.M. (2019). Conceptualising the DAIS: Implications of the "Digitalisation of Agricultural Innovation Systems" on technology and policy at multiple levels. *NJAS: Wageningen Journal of Life Sciences*, 90–91(3), 1–11 [Online]. Available at: http://www.sciencedirect.com/science/article/pii/S1573521418301532 [Accessed 27 June 2019].

Fielke, S.J., Taylor, B., Jakku, E. (2020). Digitalisation of agricultural knowledge and advice networks: A state-of-the-art review. *Agricultural Systems*, 180, 102763.

Fielke, S.J., Taylor, B.M., Jakku, E., Mooij, M., Stitzlein, C., Fleming, A., Thorburn, P.J., Webster, A.J., Davis, A., Vilas, M.P. (2021). Grasping at digitalisation: Turning imagination into fact in the sugarcane farming community. *Sustain Sci*, 16, 677–690.

Filippi, M., Frey, O. (2015). Le conseiller, une pièce maîtresse sur l'échiquier de la coopérative agricole. *Revue d'études en agriculture et environnement – Review of agricultural and environmental studies*, 96, 439–466.

Flamant, J.-C. (2010). Une brève histoire des transformations de l'agriculture au 20è siècle. *Mission Agrobiosciences*, 1–20.

Fleming, A., Jakku, E., Lim-Camacho, L., Taylor, B., Thorburn, P. (2018). Is big data for big farming or for everyone? Perceptions in the Australian grains industry. *Agronomy for Sustainable Development*, 38(3), 24.

Flichy, P. (2004). L'individualisme connecté entre la technique numérique et la société. *Reseaux*, 124(2), 17–51.

Forest, J. (2014). Petite histoire des modèles d'innovation. In *Principes d'économie de l'innovation*, Boutillier, S., Forest, J., Gallaud, D., Laperche, B., Tanguy, C., Temri, L. (eds). Peter Lang AG, Brussels, 45–57.

Fouilleux, È., Goulet, F. (2012). Firmes et developpement durable : le nouvel esprit du productivisme. *Études rurales*, 190, 131–146.

Francis, C., Jones, A., Crookston, K., Wittler, K., Goodman, S. (1986). Strip cropping corn and grain legumes: A review. *American Journal of Alternative Agriculture*, 1(4), 159–164.

Francis, C., Lieblein, G., Gliessman, S., Breland, T.A., Creamer, N., Harwood, R., Salomonsson, L., Helenius, J., Rickerl, D., Salvador, R. et al. (2003). Agroecology: The ecology of food systems. *Journal of Sustainable Agriculture*, 22, 99–118.

Frank, A.G., Mendes, G.H.S., Ayala, N.F., Ghezzi, A. (2019). Servitization and Industry 4.0 convergence in the digital transformation of product firms: A business model innovation perspective. *Technological Forecasting and Social Change*, 141, 341–351.

Fraser, A. (2019). Land grab/data grab: Precision agriculture and its new horizons. *Journal of Peasant Studies*, 46(5), 893–912.

Fraser, A. (2020). The digital revolution, data curation, and the new dynamics of food sovereignty construction. *The Journal of Peasant Studies*, 47(1), 208–226.

Fraser, A. (2021). "You can't eat data"?: Moving beyond the misconfigured innovations of smart farming. *Journal of Rural Studies*, 91, 200–207 [Online]. Available at: https://www.sciencedirect.com/science/article/pii/S0743016721001856 [Accessed 10 November 2021].

Freeman, C. (1987). *Technology Policy and Economic Performance: Lessons from Japan*. Pinter Publishers, Delfunct.

Friedman, H. and McMichael, P. (1989). Agriculture and the state system: The rise and decline of national agricultures, 1870 to the present. *Sociologia Ruralis*, 29, 93–117.

Gaitán-Cremaschi, D., Klerkx, L., Duncan, J., Trienekens, J.H., Huenchuleo, C., Dogliotti, S., Contesse, M.E., Rossing, W.A.H. (2019). Characterizing diversity of food systems in view of sustainability transitions. *Agronomy for Sustainable Development*, 39(1) [Online]. Available at: http://link.springer.com/10.1007/s13593-018-0550-2 [Accessed 28 February 2019].

Galliano, D. and Roux, P. (2006). Les inégalités spatiales dans l'usage des tic. Le cas des firmes industrielles françaises. *Revue économique*, 57(6), 1449–1475.

Gasselin, P., Lardon, S., Cerdan, C., Loudiyi, S., Sautier, D., Van der Ploeg, J.D. (2021). *Coexistence et confrontation des modèles agricoles et alimentaires*. Quæ, Versailles [Online]. Available at: https://library.oapen.org/handle/20.500.12657/47760 [Accessed 9 June 2021].

Geels, F.W. (2002). Technological transitions as evolutionary reconfiguration processes: A multi-level perspective and a case-study. *Research Policy*, 31(8), 1257–1274.

Geels, F.W. (2004). From sectoral systems of innovation to sociotechnical systems: Insights about dynamics and change from sociology and institutional theory. *Research Policy*, 33(6), 897–920.

Gerbaux, F. and Muller, P. (1984). La naissance du développement agricole en France. *Économie rurale*, 159(1), 17–22.

Giles, D.B. and Stead, V. (2021). Big Data won't feed the world: Global agribusiness, digital imperialism, and the contested promises of a new Green Revolution. *Dialectical Anthropology*, 46, 37–53. doi: 10.1007/s10624-021-09631-8 [Accessed 2 September 2021].

Giotitsas, C. (ed.) (2019). Open source agriculture: A social movement? In *Open Source Agriculture: Grassroots Technology in the Digital Era*. Springer, Berlin. doi: 10.1007/978-3-030-29341-3_4.

Gkisakis, V.D., Damianakis, K. (2020). Digital innovations for the agroecological transition: A user innovation and commons-based approach. *Journal of Sustainable and Organic Agricultural Systems*, 70, 1–4.

Glaeser, B. (2010). *The Green Revolution Revisited: Critique and Alternatives*. Taylor & Francis, New York.

Gliessman, S.R. (ed.) (1998). Agroecology: Researching the ecological basis for sustainable agriculture. In *Agroecology: Researching the Ecological Basis for Sustainable Agriculture*. Springer, New York. doi: 10.1007/978-1-4612-3252-0_1.

Glover, D., Sumberg, J., Ton, G., Andersson, J., Badstue, L. (2019). Rethinking technological change in smallholder agriculture. Report, Outlook on Agriculture.

Godard, O., Salles, J.M. (1991). Entre nature et société, les jeux de l'irréversibilité dans la construction économique et sociale du champ de l'environnement. In *Les Figures de l'irréversibilité en économie*, Boyer, R., Chavance, B., Godard, O. (eds). EHESS, Paris, 233–272.

Goulet, F. and Vinck, D. (2012). L'innovation par retrait. Contribution à une sociologie du détachement. *Revue française de sociologie*, 53(2), 195–224.

Granstrand, O. (2000). *Corporate Innovation Systems: A Comparative Study of MultiTechnology Corportations in Japan, Sweden and the USA*. Chalmers University, Gothenburg.

Gray, M.L. and Suri, S. (2019). *Ghost Work: How to Stop Silicon Valley from Building a New Global Underclass*, Houghton Mifflin Harcourt, Boston.

Gregoire, A. (2002). The mental health of farmers. *Occupational Medicine*, 52(8), 471–476.

Grenier, G. (2018). De la révolution numérique à la révolution agronomique. *Agronomie, environnement et sociétés*, 8(1), 23–32.

Griliches, Z. (1979). Issues in assessing the contribution of research and development to productivity growth. *The Bell Journal of Economics*, 10, 92–116.

Gruszka, K., Böhm, M. (2020). Out of sight, out of mind? (In)visibility of/in platform-mediated work. *New Media & Society*, 24, 1852–1871.

Gueslin, A. (1988). Crédit agricole et agriculture en France au XXe siècle. *Économie rurale*, 184(1), 107–115.

Guichard, L., Dedieu, F., Jeuffroy, M.-H., Meynard, J.M., Reau, R., Savini, I. (2017). Le plan Ecophyto de réduction d'usage des pesticides en France : décryptage d'un échec et raisons d'espérer. *Cahiers agricultures*, 26(1), 1.

Gurstein, M. (2007). Effective use: A community informatics strategy beyond the digital divide. *First Monday* [Online]. Available at: https://firstmonday.org/ojs/index.php/fm/article/view/1798 [Accessed 19 March 2021].

Haas, P.M. (1992). Introduction: Epistemic communities and international policy coordination. *International Organization*, 46(1), 1–35.

Hall, A. (2006). Public-private sector partnerships in an agricultural system of innovation: Concepts and challenges. *International Journal of Technology Management & Sustainable Development*, 5(1), 3–20.

Hall, P. and Soskice, D. (2002). *Varieties of Capitalisms*. Oxford University Press, Oxford.

Hansen, B.G. and Stræte, E.P. (2020). Dairy farmers' job satisfaction and the influence of automatic milking systems. *NJAS: Wageningen Journal of Life Sciences*, 92(1), 1–13.

Hayami, Y. and Ruttan, V.W. (1971). *Agricultural Development: An International Perspective.* The Johns Hopkins Press, London [Online]. Available at: https://www.cabdirect.org/cabdirect/abstract/19721890134 [Accessed 19 March 2020].

Heintz, W. (1990). Stratégie des organismes de collecte et qualité : le cas du blé. *Économie rurale*, 198(1), 39–39.

Hekkert, M.P., Suurs, R.A.A., Negro, S.O., Kuhlmann, S., Smits, R.E.H.M. (2007). Functions of innovation systems: A new approach for analysing technological change. *Technological Forecasting and Social Change*, 74(4), 413–432.

Hekkert, M.P., Janssen, M.J., Wesseling, J.H., Negro, S.O. (2020). Mission-oriented innovation systems. *Environmental Innovation and Societal Transitions*, 34, 76–79.

Hennessy, T., Läpple, D., Moran, B. (2016). The digital divide in farming: A problem of access or engagement? *Applied Economic Perspectives and Policy*, 38(3), 474–491.

Hervieu, B., Purseigle, F. (2009). Pour une sociologie des mondes agricoles dans la globalisation. *Études rurales*, 183, 177–200.

Higgins, V., Bryant, M., Howell, A., Battersby, J. (2017). Ordering adoption: Materiality, knowledge and farmer engagement with precision agriculture technologies. *Journal of Rural Studies*, 55, 193–202.

Hill, S.B. and MacRae, R.J. (1996). Conceptual framework for the transition from conventional to sustainable agriculture. *Journal of Sustainable Agriculture*, 7(1), 81–87.

von Hippel, E. (2006). *Democratizing Innovation*. The MIT Press, Cambridge.

Horlings, L.G. and Marsden, T.K. (2011). Towards the real green revolution? Exploring the conceptual dimensions of a new ecological modernisation of agriculture that could "feed the world". *Global Environmental Change*, 21(2), 441–452.

Hostiou, N., Fagon, J., Chauvat, S., Turlot, A., Kling, F., Boivin, X., Allain, C. (2015). Conséquences de l'élevage de précision sur le travail, les compétences et les relations homme-animal. *Elevage de précision* [Online]. Available at: https://hal.archives-ouvertes.fr/hal-01535365 [Accessed 23 April 2019].

Hostiou, N., Fagon, J., Chauvat, S., Turlot, A., Kling-Eveillard, F., Boivin, X., Allain, C. (2017). Impact of precision livestock farming on work and human-animal interactions on dairy farms. A review. *Biotechnologie, agronomie, société et environnement*, 21(4), 268–275.

Howells, J. (2006). Intermediation and the role of intermediaries in innovation. *Research Policy*, 35(5), 715–728.

Huguet, T. (2017). The connected society: Contribution to the sociological analyses between technology and society throught the example of the digital media tools. PhD Thesis, Université Paul Valéry, Montpellier [Online]. Available at: https://tel.archives-ouvertes.fr/tel-01582952 [Accessed 17 July 2020].

Huws, U. (1999). Material world: The myth of the weightless economy. *Socialist Register*, 35.

Ingram, J. (2018). Agricultural transition: Niche and regime knowledge systems' boundary dynamics. *Environmental Innovation and Societal Transitions*, 26, 117–135.

Ingram, J. and Maye, D. (2020). What are the implications of digitalisation for agricultural knowledge? *Frontiers in Sustainable Food Systems*, 4 [Online]. Available at: https://www.frontiersin.org/articles/10.3389/fsufs.2020.00066/full [Accessed 27 May 2020].

Ingram, J., Maye, D., Bailye, C., Barnes, A., Bear, C., Bell, M., Cutress, D., Davies, L., de Boon, A., Dinnie, L. et al. (2022). What are the priority research questions for digital agriculture? *Land Use Policy*, 114, 105962.

INRAE (2020). Partageons la science et l'innovation pour un avenir durable. Report, INRAE.

Interbio Occitanie (2018). L'observatoire régional de l'agriculture biologique d'Occitanie. Grandes cultures. Report [Online]. Available at: https://www.interbio-occitanie.com/content/uploads/2019/10/fiche_filiere_gc_2018_vd.pdf.

Internationale Cooperative Alliance (ICA) (1995). What is a cooperative? [Online]. Available at: https://ica.coop/en/cooperatives/what-is-a-cooperative [Accessed 4 April 2022].

Isaac, H. and Pouyat, M. (2015). Les défis de l'agriculture connectée dans une société numérique. Report, Renaissance Numérique, Paris [Online]. Available at: https://www.renaissancenumerique.org/system/attach_files/files/000/000/010/original/LB_AGRI_HD.pdf?1485335906.

Jakku, E. and Thorburn, P.J. (2010). A conceptual framework for guiding the participatory development of agricultural decision support systems. *Agricultural Systems*, 103(9), 675–682.

Jakku, E., Taylor, B.M., Fleming, A., Mason, C.M., Thorburn, P. (2016). Big data, trust and collaboration: Exploring the sociotechnical enabling conditions for big data in the grains industry. Research Report, CSIRO, EP164134.

Jakku, E., Taylor, B., Fleming, A., Mason, C., Fielke, S., Sounness, C., Thorburn, P. (2019). "If they don't tell us what they do with it, why would we trust them?" Trust, transparency and benefit-sharing in Smart Farming. *NJAS: Wageningen Journal of Life Sciences*, 90–91, 100285.

Jas, N. (2001). *Au carrefour de la chimie et de l'agriculture : les sciences agronomiques en France et en Allemagne, 1850-1914*. EAC, Paris [Online]. Available at: https://cadmus.eui.eu//handle/1814/21669 [Accessed 26 August 2020].

Javelle, A. and Garcia, F. (2022). Plantes et TIC : la médiatisation numérique du végétal permet-elle le développement de pratiques agroécologiques ? In *Agriculture et numérique. Digitalisation des systèmes agri-alimentaires. Perspectives SHS*. Société française d'économie rurale, Montpellier.

Jeanneaux, P. (2018). Agriculture numérique : quelles conséquences sur l'autonomie de la décision des agriculteurs ? *Agronomie, environnement & sociétés*, 8(1), 13.

Joly, P.B. (2010). On the economics of techno-scientific promises. In *Débordements. Mélanges offerts à Michel Callon*, Akrich, M., Barthe, Y., Muniesa, F., Mustar, P. (eds). Presses des Mines, Paris.

Joly, P.-B. (2015). Le régime des promesses technoscientifique. In *Pourquoi tant de promesses*, Audétat, M. (ed.). Hermann, Paris [Online]. Available at: https://www.researchgate.net/publication/297622208_Le_regime_des_promesses_technoscientifique.

Kernecker, M., Knierim, A., Wurbs, A., Kraus, T., Borges, F. (2020). Experience versus expectation: Farmers' perceptions of smart farming technologies for cropping systems across Europe. *Precision Agriculture*, 21(1), 34–50.

Kinsella, J. (2018). Acknowledging hard to reach farmers: Cases from Ireland. *International Journal of Agricultural Extension*, 2018, 61–69.

Kirat, T. (1991). Processus d'innovation et création d'irréversibilités : essai sur les trajectoires technologiques, le cas du machinisme agricole. PhD Thesis, Université Lumière Lyon 2, Lyon.

Kirsch, A., Kroll, J.-C., Trouvé, A. (2017). Aides directes et environnement : la politique agricole commune en question. *Économie rurale. Agricultures, alimentations, territoires*, 359, 121–139.

Kitchin, R. (2014). Big Data, new epistemologies and paradigm shifts. *Big Data & Society*, 1(1), 2053951714528481.

Kivimaa, P., Boon, W., Hyysalo, S., Klerkx, L. (2019). Towards a typology of intermediaries in sustainability transitions: A systematic review and a research agenda. *Research Policy*, 48(4), 1062–1075.

Klerkx, L. and Begemann, S. (2020). Supporting food systems transformation: The what, why, who, where and how of mission-oriented agricultural innovation systems. *Agricultural Systems*, 184, 102901.

Klerkx, L. and Leeuwis, C. (2008). Matching demand and supply in the agricultural knowledge infrastructure: Experiences with innovation intermediaries. *Food Policy*, 33(3), 260–276.

Klerkx, L. and Leeuwis, C. (2009a). Establishment and embedding of innovation brokers at different innovation system levels: Insights from the Dutch agricultural sector. *Technological Forecasting and Social Change*, 76(6), 849–860.

Klerkx, L. and Leeuwis, C. (2009b). Shaping collective functions in privatized agricultural knowledge and information systems: The positioning and embedding of a network broker in the Dutch dairy sector. *The Journal of Agricultural Education and Extension*, 15(1), 81–105.

Klerkx, L. and Rose, D. (2020). Dealing with the game-changing technologies of Agriculture 4.0: How do we manage diversity and responsibility in food system transition pathways? *Global Food Security*, 24, 100347.

Klerkx, L. and Villalobos, P. (2024). Are AgriFoodTech startups the new drivers of food systems transformation? An overview of the state of the art and a research agenda. *Global Food Security*, 40, 100726.

Klerkx, L., Aarts, N., Leeuwis, C. (2010). Adaptive management in agricultural innovation systems: The interactions between innovation networks and their environment. *Agricultural Systems*, 103(6), 390–400.

Klerkx, L., Van Mierlo, B., Leeuwis, C. (2012). Evolution of systems approaches to agricultural innovation: Concepts, analysis and interventions. In *Farming Systems Research into the 21st Century: The New Dynamic*, Darnhofer, I., Gibbon, D., Dedieu, B. (eds). Springer, Dordrecht [Online]. Available at: http://link.springer.com/10.1007/978-94-007-4503-2_20 [Accessed 5 December 2018].

Klerkx, L., Jakku, E., Labarthe, P. (2019). A review of social science on digital agriculture, smart farming and agriculture 4.0: New contributions and a future research agenda. *NJAS: Wageningen Journal of Life Sciences*, 90–91, 100315.

Kline, S.J. and Rosenberg, N. (1986). An overview of innovation. The positive sum strategy: Harnessing technology for economic growth. *The National Academy of Science*, 35.

Knierim, A., Kernecker, M., Erdle, K., Kraus, T., Borges, F., Wurbs, A. (2019). Smart farming technology innovations – Insights and reflections from the German Smart-AKIS hub. *NJAS: Wageningen Journal of Life Sciences*, 90–91, 100314.

Konrad, M.T., Nielsen, H.Ø., Pedersen, A.B., Elofsson, K. (2019). Drivers of farmers' investments in nutrient abatement technologies in five Baltic Sea countries. *Ecological Economics*, 159, 91–100.

Koschatzky, K., Baier, E., Kroll, H., Stahlecker, T. (2009). The spatial multidimensionality of sectoral innovation: The case of information and communication technologies. Report, Arbeitspapiere Unternehmen und Region [Online]. Available at: https://www.econstor.eu/handle/10419/29327 [Accessed 3 December 2018].

Koutsos, T., Menexes, G. (2017). Benefits from the adoption of precision agriculture technologies. A systematic review. In *18th Panhellenic Forestry Congress & International Workshop*, Edessa, 1–12.

Kukk, P., Moors, E.H.M., Hekkert, M.P. (2016). Institutional power play in innovation systems: The case of Herceptin®. *Research Policy*, 45(8), 1558–1569.

Labarthe, P. (2006). La privatisation du conseil agricole en question. Evolutions institutionnelles et performances des services de conseil dans trois pays européens (Allemagne, France, Pays-Bas). PhD Thesis, Université Paris Est, Créteil [Online]. Available at: https://hal.inrae.fr/tel-02819175 [Accessed 17 March 2022].

Labarthe, P. (2009). Extension services and multifunctional agriculture. Lessons learnt from the French and Dutch contexts and approaches. *Journal of Environmental Management*, 90, S193–S202.

Labarthe, P. (2010). Services immatériels et verrouillage technologique. Le cas du conseil technique aux agriculteurs. *Économies et sociétés*, 44(2), 25.

Labarthe, P. and Laurent, C. (2013). Privatization of agricultural extension services in the EU: Towards a lack of adequate knowledge for small-scale farms? *Food Policy*, 38, 240–252.

Labarthe, P. and Touzard, J.M. (2016). Regulation theory and transformation of agriculture: A literature review. *Revue de la régulation* [Online]. Available at: http://journals.openedition.org/regulation/12094 [Accessed 5 March 2019].

Labarthe, P., Prager, K., Leloup, H., Elzen, B., Collins, K., Laurant, C., Redman, M., Schoorlemmer, H.B., Sutherland, L.-A., Micheloni, C. et al. (2021a). Strengthening farm advice for innovation and sustainability. Policy Recommendations Report, 5.7.

Labarthe, P., Coléno, F., Enjalbert, J., Fugeray-Scarbel, A., Hannachi, M., Lemarié, S. (2021b). Exploration, exploitation and environmental innovation in agriculture. The case of variety mixture in France and Denmark. *Technological Forecasting and Social Change*, 172, 121028.

Labrousse, A. (2006). Éléments pour un institutionnalisme méthodologique : autonomie, variation d'échelle, réflexivité et abduction. *Économie et institutions*, 8, 5–53.

Lacoste, M., Cook, S., McNee, M., Gale, D., Ingram, J., Bellon-Maurel, V., MacMillan, T., Sylvester-Bradley, R., Kindred, D., Bramley, R. et al. (2021). On-Farm Experimentation to transform global agriculture. *Nat Food*, 3, 11–18.

Lajoie-O'Malley, A., Bronson, K., van der Burg, S., Klerkx, L. (2020). The future(s) of digital agriculture and sustainable food systems: An analysis of high-level policy documents. *Ecosystem Services*, 45, 101183.

Lamine, C. (2011). Transition pathways towards a robust ecologization of agriculture and the need for system redesign. Cases from organic farming and IPM. *Journal of Rural Studies*, 27(2), 209–219.

Lamine, C. (2012). "Changer de système" : une analyse des transitions vers l'agriculture biologique à l'échelle des systèmes agri-alimentaires territoriaux. *Terrains & travaux*, 20(1), 139–156.

Lamine, C., Meynard, J.-M., Perrot, N., Bellon (2009). Analyse des formes de transition vers des agricultures plus écologiques : les cas de l'agriculture biologique et de la protection intégrée. *Innovations agronomiques*, 4, 483–493.

Lange, S., Pohl, J., Santarius, T. (2020). Digitalization and energy consumption. Does ICT reduce energy demand? *Ecological Economics*, 176, 106760.

Larrère, R. (2002). Agriculture : artificialisation ou manipulation de la nature ? Report.

Larrère, R. and Vermersch, D. (2000). Agriculture et environnement. L'économie rurale revisitée. *Économie rurale*, 255(1), 104–113.

Latour, B. (1995). Moderniser ou écologiser. A la recherche de la Septième Cité. *Écologie & politique : sciences, culture, société*, 13, 5–27.

Laurent, C.C. and Rémy, J. (2000). L'exploitation agricole en perspective. *Le Courrier de l'environnement de l'INRA*, 41, 5–22.

Laurent, C., Maxime, F., Mazé, A., Tichit, M. (2003). Multifonctionnalité de l'agriculture et modèles de l'exploitation agricole. *Économie rurale*, 273(1), 134–152.

Laurent, C., Nguyen, G., Triboulet, P., Ansaloni, M., Bechtet, N., Labarthe, P. (2021). Institutional continuity and hidden changes in farm advisory services provision: Evidence from farmers' microAKIS observations in France. *The Journal of Agricultural Education and Extension*, 1–24.

Lawson, T. (1989). Abstraction, tendencies and stylised facts: A realist approach to economic analysis. *Cambridge Journal of Economics*, 13(1), 59–78.

Le Velly, R. and Dufeu, I. (2016). Alternative food networks as "market agencements": Exploring their multiple hybridities. *Journal of Rural Studies*, 43, 173–182.

Lebrun, M., Hostiou, N., Nowak, B., Jeanneaux, P. (2020). Agriculture de précision et transition agroécologique des exploitations : regards croisés des acteurs du monde agricole. PhD Thesis, AgroCampus Ouest, Bordeaux [Online]. Available at: https://hal.inrae.fr/hal-03280860 [Accessed 20 April 2022].

Lele, U. and Goswami, S. (2017). The fourth industrial revolution, agricultural and rural innovation, and implications for public policy and investments: A case of India. *Agricultural Economics*, 48(S1), 87–100.

Lely (2024). Alimentation automatique : meilleure efficacité alimentaire. Lely [Online]. Available at: https://www.lely.com/fr/solutions/alimentation/ [Accessed 24 October 2024].

Lenain, R., Peyrache, J., Savary, A., Severac, G. (2021). *Agricultural Robotics: Part of the New Deal? FIRA 2020 Conclusions*. Quæ, Versailles [Online]. Available at: https://www.quae-open.com/produit/171/9782759233823/agricultural-robotics-part-of-the-new-deal-fira-2020-conclusions [Accessed 28 September 2021].

van Lente, H., Boon, W.P.C., Klerkx, L. (2020). Positioning of systemic intermediaries in sustainability transitions: Between storylines and speech acts. *Environmental Innovation and Societal Transitions*, 36, 485–497.

Leveau, L., Bénel, A., Cahier, J.-P., Pinet, F., Salembier, P., Soulignac, V., Bergez, J.-E. (2019). Information and communication technology (ICT) and the agroecological transition. In *Agroecological Transitions: From Theory to Practice in Local Participatory Design*, Bergez, J.-E., Audouin, E., Therond, O. (eds). Springer, Berlin.

Lezoche, M., Hernandez, J.E., del Alemany Díaz, M.E., Panetto, H., Kacprzyk, J. (2020). Agri-food 4.0: A survey of the supply chains and technologies for the future agriculture. *Computers in Industry*, 117, 103187.

Lioutas, E.D. and Charatsari, C. (2020). Big data in agriculture: Does the new oil lead to sustainability? *Geoforum*, 109, 1–3.

Lioutas, E.D., Charatsari, C., La Rocca, G., De Rosa, M. (2019). Key questions on the use of big data in farming: An activity theory approach. *NJAS: Wageningen Journal of Life Sciences*, 90–91, 100297.

Lowenberg-DeBoer, J. and Erickson, B. (2019). Setting the record straight on precision agriculture adoption. *Agronomy Journal*, 111(4), 1552–1569.

Lucas, V. (2021). A "silent" agroecology: The significance of unrecognized sociotechnical changes made by French farmers. *Review of Agricultural, Food and Environmental Studies*. doi: 10.1007/s41130-021-00140-4.

Lucas, V., Gasselin, P. (2018). Gagner en autonomie grâce à la Cuma. Expériences d'éleveurs laitiers français à l'ère de la dérégulation et de l'agroécologie. Économie rurale. *Agricultures, alimentations, territoires*, 364, 73–89.

Lundvall, B.-Å. and Rikap, C. (2022). China's catching-up in artificial intelligence seen as a co-evolution of corporate and national innovation systems. *Research Policy*, 51(1), 104395.

Lundvall, B.-Å., Johnson, B., Andersen, E.S., Dalum, B. (2002). National systems of production, innovation and competence building. *Research Policy*, 31(2), 213–231.

Lupton, D. (2014). Beyond techno-utopia: Critical approaches to digital health technologies. *Societies*, 4(4), 706–711.

Madsen, A.K., Flyverbom, M., Hilbert, M., Ruppert, E. (2016). Big data: Issues for an international political sociology of data practices 1. *International Political Sociology*, 10(3), 275–296.

Magnin, L. (2019). La politique agricole commune et les données retardataires. *Techniques culture*, 72(2), 130–143.

Magrini, M.-B. (2018). Economie de l'innovation et des transitions : analyser et accompagner. Comprendre les processus sociotechniques de l'innovation et des transitions pour accompagner la transition agroécologique et nutritionnelle. PhD Thesis, Université de Toulouse, Toulouse.

Magrini, M.-B., Anton, M., Cholez, C., Corre-Hellou, G., Duc, G., Jeuffroy, M.-H., Meynard, J.-M., Pelzer, E., Voisin, A.-S., Walrand, S. (2016). Why are grain-legumes rarely present in cropping systems despite their environmental and nutritional benefits? Analyzing lock-in in the French agrifood system. *Ecological Economics*, 126, 152–162.

Malerba, F. (2002). Sectoral systems of innovation and production. *Research Policy*, 31(2), 247–264.

Malerba, F. (2004). *Sectoral Systems of Innovation: Concepts, Issues and Analyses of Six Major Sectors in Europe*. Cambridge University Press, Cambridge.

Markard, J. and Truffer, B. (2008). Technological innovation systems and the multi-level perspective: Towards an integrated framework. *Research Policy*, 37(4), 596–615.

Marquet, C. (2019). Binaire béton : Quand les infrastructures numériques aménagent la ville. PhD Thesis, Université Paris-Saclay, Paris [Online]. Available at: https://pastel.archives-ouvertes.fr/tel-02486197 [Accessed 16 September 2021].

Martin, B.R. (2012). The evolution of science policy and innovation studies. *Research Policy*, 41(7), 1219–1239.

Martin, A. (2019). L'agriculture veut combler son retard digital. *L'Echo* [Online]. Available at: https://www.lecho.be/dossier/pme-wallonie/l-agriculture-veut-combler-son-retard-digital/10110760.html [Accessed 20 September 2021].

Martin, T. and Schnebelin, E. (2024). Agriculture numérique : une promesse au service d'un nouvel esprit du productivisme. *Natures Sciences Sociétés* [Online]. Available at: https://www.nss-journal.org/articles/nss/abs/first/nss230046/nss230046.html [Accessed 5 February 2024].

Martin, T., Gasselin, P., Hostiou, N., Feron, G., Laurens, L., Purseigle, F. (2021). Robots and transformations of work on farms: A systematic review. In *The 2nd International Symposium on Work in Agriculture*, Clermont-Ferrand.

Mayer-Schönberger, V. and Cukier, K. (2013). *Big Data: A Revolution That Will Transform How We Live, Work, and Think*. Houghton Mifflin Harcourt, Boston.

Mazaud, C. (2017). "À chacun son métier", les agriculteurs face à l'offre numérique. *Sociologies pratiques*, 34, 39–47.

Mazoyer, M. and Roudart, L. (2017). *Histoire des agricultures du monde. Du néolithique à la crise contemporaine*. Média Diffusion, Paris.

Mazzucato, M. (2011). The entrepreneurial state. *Soundings*, 49(49), 131–142.

Mazzucato, M. (2015). Building the entrepreneurial state: A new framework for envisioning and evaluating a mission-oriented public sector. Report, Social Science Research Network [Online]. Available at: https://papers.ssrn.com/abstract=2544707 [Accessed 2 May 2022].

Mesnel, B. (2017). Les agriculteurs face à la papereasse. Policy feedbacks et bureaucratisation de la politique agricole commune. *Gouvernement et action publique*, 6(1), 33–60.

Mesnel, B. (2018). Socialiser à la biodiversité à travers la néo-libéralisation de la PAC ? *Développement durable et territoires. Économie, géographie, politique, droit, sociologie* [Online]. Available at: https://journals.openedition.org/developpementdurable/12715 [Accessed 8 February 2022].

Meynard, J.-M., Jeuffroy, M.-H., Le Bail, M., Lefèvre, A., Magrini, M.-B., Michon, C. (2017). Designing coupled innovations for the sustainability transition of agrifood systems. *Agricultural Systems*, 157, 330–339.

Micelotta, E., Lounsbury, M., Greenwood, R. (2017). Pathways of institutional change: An integrative review and research agenda. *Journal of Management*, 43(6), 1885–1910.

Michels, M., Fecke, W., Feil, J.-H., Musshoff, O., Pigisch, J., Krone, S. (2020). Smartphone adoption and use in agriculture: Empirical evidence from Germany. *Precision Agriculture*, 21(2), 403–425.

Microsoft (2021). Nourrir 10 milliards d'individus en 2050 : le grand pari de l'IoT agricole [Online]. Available at: https://experiences.microsoft.fr/articles/iot/nourrir-10-milliards-dindividu-en-2050-le-grand-pari-de-liot-agricole/ [Accessed 7 September 2021].

Microsoft (2024). FarmBeats: AI, edge & IoT for agriculture [Online]. Available at: https://www.microsoft.com/en-us/research/project/farmbeats-iot-agriculture/ [Accessed 24 October 2024].

Migliorini, P., Barberi, P., Bellon, S., Gaifami, T., Gkisakis, V.D., Peeters, A., Wezel, A. (2020). Controversial topics in agroecology: A European perspective. *International Journal of Agriculture and Natural Resources*, 47(3), 159–173.

Millerand, F. (1998). Usages des NTIC : les approches de la diffusion, de l'innovation et de l'appropriation. *Commposite*, 98(1), 19.

Ministère de l'agriculture (2021). Une délégation ministérielle pour développer le numérique et la donnée [Online]. Available at: https://agriculture.gouv.fr/une-delegation-ministerielle-pour-developper-le-numerique-et-la-donnee [Accessed 11 August 2021].

Ministère de l'agriculture et de l'alimentation (2016). alim'agri : la révolution numérique [Online]. Available at: http://agriculture.gouv.fr/alimagri-la-revolution-numerique [Accessed 9 November 2018].

Ministère de la transition écologique et ministère de l'économie (2021). Numérique et environnement : faisons converger les transitions. Report, Paris [Online]. Available at: https://www.gouvernement.fr/numerique-et-environnement-la-feuille-de-route-du-gouvernement.

Montalban, M., Frigant, V., Jullien, B. (2019). Platform economy as a new form of capitalism: A Régulationist research programme. *Cambridge Journal of Economics*, 43(4), 805–824.

Moreiro, L. (2017). Appropriation de technologies et développement durable : l'exemple de la viticulture de précision. *Innovations*, 54(3), 97–122.

Morel, K., Revoyron, E., Cristobal, M., Baret, P. (2020). Innovating within or outside dominant food systems? Different challenges for contrasting crop diversification strategies in Europe. *PLOS ONE*, 15, e0229910.

Morgan, K. and Murdoch, J. (2000). Organic vs. conventional agriculture: Knowledge, power and innovation in the food chain. *Geoforum*, 31(2), 159–173.

Morgan, K., Marsden, T., Murdoch, J. (2008). *Worlds of Food: Place, Power, and Provenance in the Food Chain*. Oxford University Press, Oxford.

Mormont, M. (2013). Écologisation : entre sciences, conventions et pratiques. *Natures Sciences Sociétés*, 21(2), 159–160.

Moschitz, H. and Stolze, M. (2018). How can we make sense of smart technologies for sustainable agriculture? A discussion paper. In *13th European IFSA Symposium. Book of Abstracts* [Online]. Available at: https://orgprints.org/id/eprint/34019/ [Accessed 23 April 2021].

Mundler, P., Labarthe, P., Laurent, C. (2006). Les disparités d'accès au conseil. Le cas de la région Rhône-Alpes. *Économie rurale. Agricultures, alimentations, territoires*, 291, 26–41.

Nelson, R.R. and Winter, S.G. (1982). *An Evolutionary Theory of Economic Change*. EMS, Caen.

Newton, J.E., Nettle, R., Pryce, J.E. (2020). Farming smarter with big data: Insights from the case of Australia's national dairy herd milk recording scheme. *Agricultural Systems*, 181, 102811.

Nguyen, G. and Purseigle, F. (2012). Les exploitations agricoles à l'épreuve de la firme. *Études rurales*, 190, 99–118.

Nguyen, G., Brailly, J., Purseigle, F. (2020). Strategic outsourcing and precision agriculture: Towards a silent reorganization of agricultural production in France? In *AgEcon Search*, San Diego [Online]. Available at: https://ageconsearch.umn.edu/record/296670 [Accessed 3 February 2020].

Nicolas, F. and Valceschini, E. (1995). *Agro-alimentaire : une économie de la qualité*, Editions Quæ – INRA, Paris.

North, D.C. (1991). Institutions. *Journal of Economic Perspectives*, 5(1), 97–112.

Ollivier, G. and Bellon, S. (2013). Dynamiques paradigmatiques des agricultures écologisées dans les communautés scientifiques internationales. *Natures Sciences Sociétés*, 21(2), 166–181.

Organization for Economic Cooperation and Development (OECD). (1997). *Proposed Guidelines for Collecting and Interpreting Technological Innovation Data – The Oslo Manual*, 2nd edition. OECD, Paris.

Oui, J. (2021a). Commodifying a "good" weather data: Commercial meteorology, low-cost stations, and the global scientific infrastructure. *Science, Technology, & Human Values*, 47(1), 29–52.

Oui, J. (2021b). La précision au secours des pollutions : des technologies numériques pour écologiser le productivisme agricole. PhD Thesis, EHESS, Paris [Online]. Available at: http://www.theses.fr/s201097 [Accessed 21 January 2022].

Pavitt, K. (1984). Sectoral patterns of technical change: Towards a taxonomy and a theory. *Research Policy*, 13(6), 343–373.

Pearson, S., May, D., Leontidis, G., Swainson, M., Brewer, S., Bidaut, L., Frey, J.G., Parr, G., Maull, R., Zisman, A. (2019). Are distributed ledger technologies the panacea for food traceability? *Global Food Security*, 20, 145–149.

Peirce, C.S. (1974). *Collected Papers of Charles Sanders Peirce*. Harvard University Press, Cambridge.

Perez, C. (1983). Structural change and assimilation of new technologies in the economic and social systems. *Futures*, 15(5), 357–375.

Phillips, T., Klerkx, L., McEntee, M. (2018). An investigation of social media's roles in knowledge exchange by farmers. Report.

Pierson, P. (2000). Increasing returns, path dependence, and the study of politics. *American Political Science Review*, 94(2), 251–267.

Pigford, A.-A.E., Hickey, G.M., Klerkx, L. (2018). Beyond agricultural innovation systems? Exploring an agricultural innovation ecosystems approach for niche design and development in sustainability transitions. *Agricultural Systems*, 164, 116–121.

Pillaud, H. (2015). *Agronumericus : internet est dans le pré*. Editions France Agricole, Paris.

Piot-Lepetit, I. (2019). Quel devenir pour les startups proposant des services numériques dans une vision globale ? Exemples de fusions/acquisitions dans le secteur des intrants et machines agricoles. In *Numérique et intégration des agriculteurs dans les chaînes de valeur : Existant et devenir en amont et à l'aval*. DigitAg, Montpellier.

Piot-Lepetit, I. (2022). Start-ups and digital innovation in the agri-food sector. *Annales des mines*, 19, 97–102.

Piriou, S. (2002). L'institutionnalisation de l'agriculture biologique (1980-2000). PhD Thesis, Ecole Nationale Supérieure Agronomique de Rennes, Rennes [Online]. Available at: tel-00910272 [Accessed 15 April 2020].

Pitron, G. (2017). *La guerre des métaux rares, la face cachée de la transition énergétique*. Les Liens qui Libèrent, Paris.

Plumecocq, G., Corso, J.-P.D., Kephaliacos, C. (2015). Quel rôle pour les coopératives dans la mise en œuvre de mesures agroenvironnementales ? In *Conseil privé en agriculture*, Compagnone, C., Goulet, F., Labarthe, P. (eds). Éducagri éditions and Éditions Quæ, Dijon/Paris.

Plumecocq, G., Debril, T., Duru, M., Magrini, M.-B., Sarthou, J.P., Therond, O. (2018). The plurality of values in sustainable agriculture models: diverse lock-in and coevolution patterns. *Ecology and Society*, 23(1) [Online]. Available at: https://www.jstor.org/stable/26799066 [Accessed 15 April 2020].

Potier, N. (2021). Transition écologique : vers un levier numérique ? *Tactis* [Online]. Available at: https://www.tactis.fr/transition-ecologique/ [Accessed 23 September 2021].

Prause, L. (2021). Digital agriculture and labor: A few challenges for social sustainability. *Sustainability*, 13(11), 5980.

Prause, L., Hackfort, S., Lindgren, M. (2020). Digitalization and the third food regime. *Agriculture and Human Values*, 38, 641–655.

Regan, Á. (2019). "Smart farming" in Ireland: A risk perception study with key governance actors. *NJAS: Wageningen Journal of Life Sciences*, 90–91, 100292.

Reinert, M. (2022). L'agriculture numérique ou la fin des paysans. Reporterre, le média de l'écologie [Online]. Available at: https://reporterre.net/L-agriculture-numerique-ou-la-fin-des-paysans [Accessed 29 March 2022].

Rémy, J. (1987). La crise de professionnalisation en agriculture : les enjeux de la lutte pour le contrôle du titre d'agriculteur. *Sociologie du travail*, 29(4), 415–441.

Rénier, L., Cardona, A., Goulet, F., Ollivier, G. (2022). La proximité à distance. *Reseaux*, 231(1), 225–257.

Réseau Action Climat (2020). L'agriculture de précision : un modèle aux antipodes de la transition écologique et sociale. Report, Réseau Action Climat France [Online]. Available at: https://reseauactionclimat.org/wp-content/uploads/2020/09/notes-rac-agriculture-de-precision.pdf.

Réseau Rural Français (2018). L'impact des usages du numérique sur le développement rural : constats et questionnements. FEADER [Online]. Available at: https://pos-occitanie.fr/files/pmedia/public/r1026_9_2018_rrf_etude_animation_thematique_impact_numerique_developpement_rural.pdf

Richardson, L. and Bissell, D. (2019). Geographies of digital skill. *Geoforum*, 99, 278–286.

Rijswijk, K., Klerkx, L., Turner, J.A. (2019). Digitalisation in the New Zealand agricultural knowledge and innovation system: Initial understandings and emerging organisational responses to digital agriculture. *NJAS: Wageningen Journal of Life Sciences*, 90–91, 100313.

Rikap, C. (2021). *Capitalism, Power and Innovation. Intellectual Monopoly Capitalims Uncovered*. Routledge, London.

Ringsberg, H. (2014). Perspectives on food traceability: A systematic literature review. *Supply Chain Management: An International Journal*, 19, 558–576.

Rip, A. and Kemp, R. (1998). Technological change. In *Human Choice and Climate Change*, Raynor, S. and Malone, E. (eds). Battelle Press, Columbus, OH, 327–399.

Rogers, E.M. (1962). *Diffusion of Innovations*. Free Press, New York.

Rogers, E.M. (2010). *Diffusion of Innovations*. Simon and Schuster, London.

Rolandi, S., Brunori, G., Bacco, M., Scotti, I. (2021). The digitalization of agriculture and rural areas: Towards a taxonomy of the impacts. *Sustainability*, 13(9), 5172.

Rose, D.C., Morris, C., Lobley, M., Winter, M., Sutherland, W.J., Dicks, L.V. (2018). Exploring the spatialities of technological and user re-scripting: The case of decision support tools in UK agriculture. *Geoforum*, 89, 11–18.

Rose, D.C., Wheeler, R., Winter, M., Lobley, M., Chivers, C.-A. (2021). Agriculture 4.0: Making it work for people, production, and the planet. *Land Use Policy*, 100, 104933.

Rotz, S., Duncan, E., Small, M., Botschner, J., Dara, R., Mosby, I., Reed, M., Fraser, E.D.G. (2019a). The politics of digital agricultural technologies: A preliminary review. *Sociologia Ruralis*, 59(2), 203–229.

Rotz, S., Gravely, E., Mosby, I., Duncan, E., Finnis, E., Horgan, M., LeBlanc, J., Martin, R., Neufeld, H.T., Pant, L. et al. (2019b). Automated pastures and the digital divide: How agricultural technologies are shaping labour and rural communities. *Journal of Rural Studies* [Online]. Available at: https://linkinghub.elsevier.com/retrieve/pii/S0743016718307769 [Accessed 19 February 2019].

Ryan, M. (2019). Agricultural Big Data analytics and the ethics of power. *Journal of Agricultural & Environmental Ethics*, 33, 49–69.

Saïsset, L.-A. (2021). Tensions of agricultural co-op democratic governance. A case study in south of France. Report [Online]. Available at: https://ideas.repec.org/p/hal/journl/hal-03284348.html [Accessed 25 February 2022].

Scherer, F.M. (1983). The propensity to patent. *International Journal of Industrial Organization*, 1, 107–128.

Schmookler, J. (1966). *Invention and Economic Growth*. Harvard University Press, Cambridge.

Schnebelin, E. (2022a). Le développement du numérique dans les trajectoires d'écologisation de l'agriculture France. PhD Thesis, SupAgro, Montpellier [Online]. Available at: https://www.theses.fr/s218357 [Accessed 26 September 2022].

Schnebelin, E. (2022b). Linking the diversity of ecologisation models to farmers' digital use profiles. *Ecological Economics*, 196, 107422.

Schnebelin, E., Labarthe, P., Touzard, J.-M. (2021). How digitalisation interacts with ecologisation? Perspectives from actors of the French Agricultural Innovation System. *Journal of Rural Studies*, 86, 599–610.

Schnebelin, E., Touzard, J.-M., Labarthe, P., Macaine, I. (2022a). Quel rôle pour les coopératives dans l'écologisation et la numérisation de l'agriculture ? L'exemple de la viticulture en Occitanie (France). *RECMA*, 366(4), 33–49.

Schnebelin, E., Valiente, C., Touzard, J.-M., Labarthe, P.P. (2022b). Usages du numérique en agriculture : enquête auprès de 98 exploitations ayant des grandes cultures en région Occitanie. Report, Inrae [Online]. Available at: https://hal.inrae.fr/hal-03573156 [Accessed 21 April 2022].

Scholz, T. (ed.) (2012). *Digital Labor*. Routledge, London. [Online]. Available at: https://www.taylorfrancis.com/books/9781136506703 [Accessed 21 September 2021].

Schumpeter, J.A. (1935). The analysis of economic change. *The Review of Economics and Statistics*, 17(4), 2–10.

Séjeau, W. (2004). René Dumont agronome. *Ruralia. Sciences sociales et mondes ruraux contemporains* [Online]. Available at: https://journals.openedition.org/ruralia/1027 [Accessed 22 April 2022].

Serrano, E., Petit, A., Gaviglio, C., Gontier, L., Dufourcq, T., Cogo, R., Massol, T., Mora, J., Rafenne, X., Cichosz, B. (2019). Ecoviti Sud-Ouest : performances de systèmes viticoles innovants à faible niveau d'intrants phytopharmaceutiques sur le bassin de production Sud-Ouest. *Innovations agronomiques*, 76, 254–272.

Servolin, C. (1989). *L'agriculture moderne*. Le Seuil, Paris.

Shove, E. and Walker, G. (2007). Caution! Transitions ahead: Politics, practice, and sustainable transition management. *Environment and Planning A: Economy and Space*, 39(4), 763–770.

Sine, M., Brizot, C., Brun, F. (2018). La blockchain dans le monde agricole : quels usages pour quels bénéfices ? In *Conférence sur la blockchain en agriculture – Salon de l'agriculture 2018*, Paris.

Sippel, S.R. and Dolinga, M. (2023). Constructing agri-food for finance: Startups, venture capital and food future imaginaries. *Agriculture and Human Values*, 40(2), 475–488.

Skaalsveen, K., Ingram, J., Urquhart, J. (2020). The role of farmers' social networks in the implementation of no-till farming practices. *Agricultural Systems*, 181, 1–14.

Solow, R.M. (1956). A contribution to the theory of economic growth. *The Quarterly Journal of Economics* 70, 65–94.

Sonka, S. (2021). Digital technologies, big data, and agricultural innovation. In *The Innovation Revolution in Agriculture: A Roadmap to Value Creation*, Campos, H. (ed.). Springer, Berlin [Online]. Available at: http://link.springer.com/10.1007/978-3-030-50991-0 [Accessed 4 March 2022].

Sonnino, R. and Marsden, T. (2006). Beyond the divide: Rethinking relationships between alternative and conventional food networks in Europe. *Journal of Economic Geography*, 6(2), 181–199.

Spaargaren, G. and Mol, A.P.J. (1992). Sociology, environment, and modernity: Ecological modernization as a theory of social change. *Society & Natural Resources*, 5(4), 323–344.

Spielman, D.J. and Birner, R. (2008). How innovative is your agriculture? Using innovation indicators and benchmarks to strengthen national agricultural innovation systems. Agriculture and Rural Development Discussion Paper no. 41, World Bank, Washington, D.C. [Online]. Available at: http://documents.worldbank.org/curated/en/696461468316131075.

Srnicek, N. (2018). *Capitalisme de plateforme : l'hégémonie de l'économie numérique.* Lux, Montréal.

Stassart, P.M. and Jamar, D. (2009). Agriculture biologique et verrouillage des systèmes de connaissances. Conventionalisation des filières agroalimentaires bio. *Innovations agronomiques*, 4, 313–328.

Stassart, P.M., Baret, P., Grégoire, J.-C., Hance, T., Mormont, M., Reheul, D., Stilmant, D., Vanloqueren, G., Visser, M. (2012). L'agroécologie : trajectoire et potentiel. Pour une transition vers des systèmes alimentaires durables. In *Agroécologie, entre pratiques et sciences sociales*, Van Dam, D., Streith, M., Nizet, J., Stassart, P. (eds). Éducagri éditions, Dijon [Online]. Available at: https://www.cairn.info/agroecologie-9782844448767-page-25.htm [Accessed 12 February 2021].

Stephan, A., Schmidt, T.S., Bening, C.R., Hoffmann, V.H. (2017). The sectoral configuration of technological innovation systems: Patterns of knowledge development and diffusion in the lithium-ion battery technology in Japan. *Research Policy*, 46(4), 709–723.

Steyaert, P., Barbier, M., Cerf, M., Levain, A., Loconto, A.M. (2016). Role of intermediation in the management of complex sociotechnical transitions. Report.

Stone, G.D. (2022). Surveillance agriculture and peasant autonomy. *Journal of Agrarian Change*, 22, 608–631.

Sulecki, J. (2018). Association seeks definitive definition of "precision agriculture". What's your vote? PrecisionAg [Online]. Available at: https://www.precisionag.com/market-watch/association-seeks-definitive-definition-of-precision-agriculture-whats-your-vote/ [Accessed 6 July 2021].

Sutherland, L.-A. (2015). New entrants into farming: Lessons to foster innovation and entrepreneurship (EIP Agriculture and Innovation). *Food and Agriculture Organization of the United Nations and International Telecommunication Union*, European Union.

Sutherland, L.-A., Madureira, L., Dirimanova, V., Bogusz, M., Kania, J., Vinohradnik, K., Creaney, R., Duckett, D., Koehnen, T., Knierim, A. (2017). New knowledge networks of small-scale farmers in Europe's periphery. *Land Use Policy*, 63, 428–439.

Sutherland, L.-A., Labarthe, P., Elzen, B., Adamsone Fiskovica, A. (2018). AgriLink's multi-level conceptual framework. Report, AgriLink Project [Online]. Available at: https://old.agrilink2020.eu/wp-content/uploads/2019/02/AgriLink-conceptual-framework.-Theory-primers.20.pdf..

Sutherland, L.-A., Madureira, L., Elzen, B., Noble, C., Bechtet, N., Townsend, L., Zarokosta, E., Triboulet, P. (2022). What can we learn from droppers and nonadopters about the role of advice in agricultural innovation? *EuroChoices*, 21(1), 40–49.

Tardieu, V. (2017). *Agriculture connectée. Arnaque ou remède ?* Belin, Paris.

Temri, L. (2014). Stratégies d'entreprises et systèmes sectoriels d'innovation. In *Principes d'économie de l'innovation*, Boutillier, S., Forest, J., Laperche, B., Tanguy, C., Termi, L. (eds). Peter Lang, Brussels.

Tenner, E. (1997). *Why Things Bite Back: Technology and the Revenge of Unintended Consequences.* Vintage, London.

Terranova, T. (2000). Free labor: Producing culture for the digital economy. *Social Text*, 18(2), 33–58.

Terre-Net (2021). Étude Agrinautes 2020 Réseau 4G, OAD, e-commerce : 5 infos sur la pratique numérique des agriculteurs [Online]. Available at: https://www.terre-net.fr/actualite-agricole/economie-social/article/les-cinq-informations-a-retenir-de-l-usage-du-numerique-chez-les-agriculteurs-202-176145.html [Accessed 13 August 2021].

Therond, O., Duru, M., Roger-Estrade, J., Richard, G. (2017). A new analytical framework of farming system and agriculture model diversities. A review. *Agronomy for Sustainable Development*, 37(3), 21.

Torre, A. and Tanguy, C. (2014). Les systèmes territoriaux d'innovation : fondements et prolongements actuels. In *Principes d'économie de l'innovation*, Boutillier, S., Forest, J., Laperche, B., Tanguy, C., Termi, L. (eds). Peter Lang, Brussels.

Torrez, C., Miller, N., Ramsey, S., Griffin, T. (2016). Factors influencing the adoption of precision agricultural technologies by Kansas farmers. Kansas State University Department of Agricultural Economics Extension Publication 9.

Touzard, J.-M. (2010). Ancrage territorial et construction de règles dans des coopératives viticoles du Midi. In *Le temps des SYAL*, Muchnik, J. and de Sainte Marie, C. (eds). Quæ, Versailles [Online]. Available at: https://www.cairn.info/le-temps-des-syal-9782759205004-page-265.htm [Accessed 4 October 2021].

Touzard, J.M. (2014). Les approches sectorielles de l'innovation. In *Principes d'économie de l'innovation,* Boutillier, S., Forest, J., Gallaud, D., Laperche, B., Tanguy, C., Temri, L. (eds). Peter Lang, Brussels, 235–245.

Touzard, J.-M. (2018). L'innovation agricole et agroalimentaire au XXIe siècle : maintien, effacement ou renouvellement de ses spécificités ? In *Innovation et développement dans les systèmes agricoles et alimentaires*, Faure, G., Chiffoleau, Y., Goulet, F., Temple, L., Touzard, J.-M. (eds). Quæ, Versailles.

Touzard, J.-M. and Draperi, J.-F. (2003). *Les coopératives entre territoires et mondialisation.* L'Harmattan, Paris [Online]. Available at: https://hal.inrae.fr/hal-03321459 [Accessed 25 February 2022].

Touzard, J.-M. and Labarthe, P. (2017). Théorie de la régulation et transformations de l'agriculture : nouveaux enjeux, nouvelles perspectives de recherche ? In *Transformations agricoles et agroalimentaires : entre écologie et capitalisme*, Allaire, G. and Daviron, B. (eds). Quæ, Versailles [Online]. Available at: https://hal.archives-ouvertes.fr/hal-01530806 [Accessed 29 March 2022].

Touzard, J.-M. and Labarthe, P. (2018). Regulation Theory and transformation of agriculture: A literature review. In *Ecology, Capitalism and the New Agricultural Economy*, Allaire, G. and Daviron, B. (eds). Routledge, Abingdon.

Touzard, J.-M., Temple, L., Faure, G., Triomphe, B. (2014). Systèmes d'innovation et communautés de connaissances dans le secteur agricole et agroalimentaire. *Innovations*, 43(1), 13.

Trouvé, A. (2009). Les régions, porteuses de nouveaux compromis pour l'agriculture ? *Revue de la régulation. Capitalisme, institutions, pouvoirs*, 11 [Online]. Available at: https://journals.openedition.org/regulation/7550 [Accessed 31 March 2022].

Tsing, A.L. (2017). *Le champignon de la fin du monde : sur la possibilité de vie dans les ruines du capitalisme.* La Découverte, Paris.

Vadén, T. (2004). Digital nominalism. Notes on the ethics of information society in view of the ontology of the digital. *Ethics and Information Technology*, 6(4), 223–231.

Valceschini, E. (2006). Valorisation économique de la traçabilité : les stratégies de qualité dans l'agroalimentaire. *Techniques d'Ingénieur*. doi: 10.51257/a-v1-tr920.

Valiorgue, B., Bargues, É.B., Hollandts, X. (2020). Quelles évolutions de la raison d'être des coopératives agricoles françaises ? Regard historique sur un construit social. *RECMA*, 358(4), 23–38.

Van Bers, C., Delaney, A., Eakin, H., Cramer, L., Purdon, M., Oberlack, C., Evans, T., Pahl-Wostl, C., Eriksen, S., Jones, L. et al. (2019). Advancing the research agenda on food systems governance and transformation. *Current Opinion in Environmental Sustainability*, 39, 94–102.

Van der Ploeg, J.D. (1994). Styles of farming: An introductory note on concepts and methodology. In *Born from Within – Practice and Perspectives of Endeogenous Rural Development*, Van der Ploeg, J.D., Long, A. (eds). Van Gorcum, Assen, 7–31.

Van der Ploeg, J.D. (2018). Differentiation: Old controversies, new insights. *The Journal of Peasant Studies*, 45(3), 489–524.

Van der Ploeg, J.D., Douwe, J., Barjolle, D., Bruil, J., Brunori, G., Madureira, L.M.C., Dessein, J., Drag, Z., Fink-Kessler, A., Gasselin, P. et al. (2019). The economic potential of agroecology: Empirical evidence from Europe. *Journal of Rural Studies*, 71, 46–61.

Van Es, H., Woodard, J. (2017). Innovation in agriculture and food systems in the Digital Age. *The Global Innovation Index*, 99, 97–104.

Van Hulst, F., Ellis, R., Prager, K., Msika, J. (2020). Using co-constructed mental models to understand stakeholder perspectives on agro-ecology. *International Journal of Agricultural Sustainability*, 0, 18(2), 172–195.

Vanloqueren, G. and Baret, P.V. (2008). Why are ecological, low-input, multi-resistant wheat cultivars slow to develop commercially? A Belgian agricultural "lock-in" case study. *Ecological Economics*, 66(2), 436–446.

Vanloqueren, G. and Baret, P.V. (2009). How agricultural research systems shape a technological regime that develops genetic engineering but locks out agroecological innovations. *Research Policy*, 38(6), 971–983.

Verdouw, C. (2016). Internet of Things in agriculture. *CAB Reviews: Perspectives in Agriculture, Veterinary Science, Nutrition and Natural Resources*, 11(035) [Online]. Available at: https://www.cabi.org/cabreviews/review/20163379897 [Accessed 20 November 2018].

Visser, O., Sippel, S.R., Thiemann, L. (2021). Imprecision farming? Examining the (in)accuracy and risks of digital agriculture. *Journal of Rural Studies*, 86, 623–632.

Walliser, B. (1989). Théorie des jeux et genèse des institutions. *Recherches économiques de Louvain*, 55(4), 339–364.

Walter, A., Finger, R., Huber, R., Buchmann, N. (2017). Opinion: Smart farming is key to developing sustainable agriculture. *Proceedings of the National Academy of Sciences*, 114(24), 6148–6150.

Warren, M. (2007). The digital vicious cycle: Links between social disadvantage and digital exclusion in rural areas. *Telecommunications Policy*, 31(6), 374–388.

Weller, J.-M. (2006). Il faut sauver l'agriculteur Poulard de la télédétection : le souci du public à l'épreuve du travail administratif. *Politiques et management public*, 24(3), 109–122.

Wenger, E. (2010). Communities of practice and social learning systems: The career of a concept. In *Social Learning Systems and Communities of Practice*, Blackmore, C. (ed.). Springer, London. doi: 10.1007/978-1-84996-133-2_11.

Wezel, A., Bellon, S., Doré, T., Francis, C., Vallod, D., David, C. (2009). Agroecology as a science, a movement and a practice. A review. *Agronomy for Sustainable Development*, 29(4), 503–515.

Williams, R. and Edge, D. (1996). The social shaping of technology. *Research Policy*, 25(6), 865–899.

Wilkinson, J. and Goodman, D. (2018). Food regime analysis: A reassessment. In *Ecology, Capitalism and the New Agricultural Economy*, Allaire, G. and Daviron, B. (eds). Routledge, New York, 142–162.

Wittman, H., James, D., Mehrabi, Z. (2020). Advancing food sovereignty through farmer-driven digital agroecology. *International Journal of Agriculture and Natural Resources*, 47(3), 235–248.

Wolf, S.A. and Buttel, F.H. (1996). The political economy of precision farming. *American Journal of Agricultural Economics*, 78(5), 1269.

Wolf, S.A. and Nowak, P.J. (1995). Development of information intensive agrichemical management services in Wisconsin. *Environmental Management*, 19(3), 371.

Wolf, S.A. and Wood, S.D. (1997). Precision farming: Environmental legitimation, commodification of information, and industrial coordination 1. *Rural Sociology*, 62(2), 180–206.

Wolfert, S., Ge, L., Verdouw, C., Bogaardt, M.-J. (2017). Big Data in smart farming. A review. *Agricultural Systems*, 153, 69–80.

Wolfert, S., Verdouw, C., van Wassenaer, L., Dolfsma, W., Klerkx, L. (2023). Digital innovation ecosystems in agri-food: Design principles and organizational framework. *Agricultural Systems*, 204, 103558.

Yang, H., Klerkx, L., Leeuwis, C. (2014). Functions and limitations of farmer cooperatives as innovation intermediaries: Findings from China. *Agricultural Systems*, 127, 115–125.

Zuboff, S. (2019). Surveillance capitalism and the challenge of collective action. *New Labor Forum*, 28(1), 10–29.

Index

F, G

H, I

J, K

L, M

O, P

R, S

T, U

V, W

Other titles from

iSTE

in

Innovation, Entrepreneurship and Management

2026

GILBERT Patrick, LE BOULAIRE Martine
The Hidden Faces of Innovation
(Technological Changes and Human Resources – Volume 6)

ORUEZABALA Gwenaëlle, ABBAD Hicham
Digital Strategies for Supply Chains: From Design to Deployment

SEBAI Jihane, SZOSTAK Bérangère L.
Social Innovation Through Design in Healthcare 1: Thinking, Understanding and Implementing
(Health and Innovation – Volume 7)

2025

CHABAUD Didier, EYNAUD Philippe, RAULET-CROSET Nathalie
Reinventing Territories through Solidarity
(Solidarity-based Management and Organizations – Volume 1)

DIMOTIKALIS Yannis, SKIADIS Christos H.
Data Analysis and Related Applications 5: Models, Methods and Techniques
(Big Data, Artificial Intelligence and Data Analysis Set – Volume 13)

GORIA Stéphane
Serious Games and Innovation Gains: Think Innovation through Games
(Innovation in Engineering and Technology – Volume 9)

GRENIER Corinne, FRANKLIN-JOHNSON Elizabeth,
CAJAIBA-SANTANA Giovany
Promoting a More Inclusive Society for Dependent or Disabled People: New Paradigms
(Health and Innovation – Volume 6)

KLESZCZOWSKI Julien, RAULET-CROSET Nathalie
Social Impact in Solidarity-based Organizations: Values and Assessment Practices
(Solidarity-based Management and Organizations – Volume 2)

2024

AÏT-EL-HADJ Smaïl
Environmental Transition and Technological Change
(Smart Innovation – Volume 43)

ALBERT-CROMARIAS Anne, ALBERTINI Thérèse, TERRAMORSI Patrice
Collective Dynamics and Territories: 9 Issues for Competitive, Innovative and Sustainable Territories
(Territorial Entrepreneurship and Innovation – Volume 3)

BRASSEUR Martine, BARTOLI Annie, CHABAUD Didier, GROUIEZ Pascal, ROUET Gilles
Inclusive Territories 2: Role of Institutions and Local Actors
(Territorial Entrepreneurship and Innovation Set – Volume 2)

COURIE-LEMEUR Aline
Innovation, Collective Intelligence and Resiliency in Healthcare Organizations
(Health and Innovation Set – Volume 4)

DARTIGUEPEYROU Carine, SALOFF-COSTE Michel
Futures in Action: Strategic Anticipations and Deployments in Organizations Facing the Future
(Innovation and Technology Set – Volume 19)

DELÉCOLLE Thierry, JACOB Florence, PRIM-ALLAZ Isabelle
New Frontiers of Customer Strategy: Managing Sustainable, Environmental and Ethical Transitions

DIMOTIKALIS Yannis, SKIADIS Christos H.
Data Analysis and Related Applications 3: Theory and Practice – New Approaches
(Big Data, Artificial Intelligence and Data Analysis Set – Volume 11)
Data Analysis and Related Applications 4: New Approaches
(Big Data, Artificial Intelligence and Data Analysis Set – Volume 12)

FLORA Luigi, GRENIER Corinne, PONISGNON Frédéric
Experience in Healthcare Innovation: Fad or New Paradigm?
(Health and Innovation – Volume 5)

MERLIN-BROGNIART Céline
Functional Economy: Economic Models, Challenges and Innovation Dynamics
(Smart Innovation – Volume 44)

PARMENTIER Guy
Organizational Creative Capabilities: Management Factors, Processes and Devices
(Organizational Creativity Set – Volume 1)

RUIZ Émilie
The Creative Capabilities of Open Organizations
(Organizational Creativity Set – Volume 2)

VERNIER Marie-France
Eco-Design and Ecological Transition: Questioning the Economic Model
(Smart Innovation Set – Volume 42)

2023

BOLLINGER Sophie
Management Control and Creativity: Challenges of Managing Innovation Processes
(Smart Innovation Set – Volume 41)

BOUVIER-PATRON Paul
Frugal Innovation and Innovative Creation
(Smart Innovation Set – Volume 40)

BRASSEUR Martine, BARTOLI Annie, CHABAUD Didier, GROUIEZ Pascal, ROUET Gilles
Inclusive Territories 1: Role of Enterprises and Organizations
(Territorial Entrepreneurship and Innovation Set – Volume 1)

CASADELLA Vanessa, UZUNIDIS Dimitri
Agri-Innovations and Development Challenges: Engineering, Value Chains and Socio-economic Models
(Innovation in Engineering and Technology Set – Volume 8)

DARTIGUEPEYROU Carine, SALOFF-COSTE Michel
Futures: The Great Turn
(Innovation and Technology Set – Volume 18)

PEYROUX Élisabeth, RAIMOND Christine, VIEL Vincent, LAVIE Émilie
Development and Territorial Restructuring in an Era of Global Change: Theories, Approaches and Future Research Perspectives

SAULAIS Pierre
Knowledge and Ideation: Inventive Knowledge Analysis for Ideation Stimulation
(Innovation and Technology Set – Volume 17)

2022

AOUINAÏT Camille
Open Innovation Strategies
(Smart Innovation Set – Volume 39)

BOUCHÉ Geneviève
Productive Economy, Contributory Economy: Governance Tools for the Third Millennium
(Innovation and Technology Set – Volume 15)

BRUYÈRE Christelle
Caring Management in Health Organizations: A Lever for Crisis Management
(Health and Innovation Set – Volume 3)

ENJOLRAS Manon
Innovation and Export: The Joint Challenge of the Small Company
(Smart Innovation Set – Volume 37)

FLEURY Sylvain, RICHIR Simon
Immersive Technologies to Accelerate Innovation: How Virtual and Augmented Reality Enables the Co-Creation of Concepts
(Smart Innovation Set – Volume 38)

HELLER David
Valuation of the Liability Structure by Real Options
(Modern Finance, Management Innovation and Economic Growth Set – Volume 5)

MATHIEU Valérie
A Customer-oriented Manager for B2B Services: Principles and Implementation

MORALES Lucía, DZEVER Sam, TAYLOR Robert
Asia-Europe Industrial Connectivity in Times of Crisis
(Innovation and Technology Set – Volume 16)

NOËL Florent, SCHMIDT Géraldine
Employability and Industrial Mutations: Between Individual Trajectories and Organizational Strategic Planning
(Technological Changes and Human Resources Set – Volume 4)

DE SAINT JULIEN Odile
The Innovation Ecosystem as a Source of Value Creation: A Value Creation Lever for Open Innovation
(Diverse and Global Perspectives on Value Creation Set – Volume 4)

SALOFF-COSTE Michel
Innovation Ecosystems: The Future of Civilizations and the Civilization of the Future
(Innovation and Technology Set – Volume 14)

SOULÉ Bastien, HALLÉ Julie, VIGNAL Bénédicte, BOUTROY Éric, NIER Olivier
Innovation in Sport: Innovation Trajectories and Process Optimization
(Smart Innovation Set – Volume 35)

VALLIER Estelle
Innovation in Clusters: Science–Industry Relationships in the Face of Forced Advancement
(Smart Innovation Set – Volume 36)

VAYRE Emilie
Digitalization of Work: New Spaces and New Working Times
(Technological Changes and Human Resources Set – Volume 5)

ZAFEIRIS Konstantinos N, SKIADIS Christos H, DIMOTIKALIS Yannis, KARAGRIGORIOU Alex, KARAGRIGORIOU-VONTA Christina
Data Analysis and Related Applications 1: Computational, Algorithmic and Applied Economic Data Analysis
(Big Data, Artificial Intelligence and Data Analysis Set – Volume 9)
Data Analysis and Related Applications 2: Multivariate, Health and Demographic Data Analysis
(Big Data, Artificial Intelligence and Data Analysis Set – Volume 10)

2021

ACH Yves-Alain, RMADI-SAÏD Sandra
Financial Information and Brand Value: Reflections, Challenges and Limitations

ARCADE Jacques
Strategic Engineering
(Innovation and Technology Set – Volume 11)

BÉRANGER Jérôme, RIZOULIÈRES Roland
The Digital Revolution in Health
(Health and Innovation Set – Volume 2)

BOBILLIER CHAUMON Marc-Eric
Digital Transformations in the Challenge of Activity and Work: Understanding and Supporting Technological Changes
(Technological Changes and Human Resources Set – Volume 3)

BUCLET Nicolas
Territorial Ecology and Socio-ecological Transition
(Smart Innovation Set – Volume 34)

DIMOTIKALIS Yannis, KARAGRIGORIOU Alex, PARPOULA Christina, SKIADIS Christos H
Applied Modeling Techniques and Data Analysis 1: Computational Data Analysis Methods and Tools
(Big Data, Artificial Intelligence and Data Analysis Set - Volume 7)
Applied Modeling Techniques and Data Analysis 2: Financial, Demographic, Stochastic and Statistical Models and Methods
(Big Data, Artificial Intelligence and Data Analysis Set – Volume 8)

DISPAS Christophe, KAYANAKIS Georges, SERVEL Nicolas, STRIUKOVA Ludmila
Innovation and Financial Markets
(Innovation between Risk and Reward Set – Volume 7)

GAUDIN Thierry, MAUREL Marie-Christine, POMEROL Jean-Charles
Chance, Calculation and Life

GASMI Nacer
Corporate Innovation Strategies: Corporate Social Responsibility and Shared Value Creation
(Smart Innovation Set – Volume 33)

GIORGINI Pierre
The Contributory Revolution
(Innovation and Technology Set – Volume 13)

GOGLIN Christian
Emotions and Values in Equity Crowdfunding Investment Choices 1: Transdisciplinary Theoretical Approach
Emotions and Values in Equity Crowdfunding Investment Choices 2: Modeling and Empirical Study

GRENIER Corinne, OIRY Ewan
Altering Frontiers: Organizational Innovations in Healthcare
(Health and Innovation Set – Volume 1)

GUERRIER Claudine
Security and Its Challenges in the 21st Century
(Innovation and Technology Set – Volume 12)

HELLER David
Performance of Valuation Methods in Financial Transactions (Modern Finance, Management Innovation and Economic Growth Set – Volume 4)

LEHMANN Paul-Jacques
Liberalism and Capitalism Today

LIMA Marcos
Entrepreneurship and Innovation Education: Frameworks and Tools (Smart Innovation Set – Volume 32)

MACHADO Carolina, DAVIM J. Paulo
Sustainable Management for Managers and Engineers

MEUNIER François-Xavier
Dual Innovation Systems: Concepts, Tools and Methods (Smart Innovation Set – Volume 31)

UZUNIDIS Dimitri, KASMI Fedoua, ADATTO Laurent
Innovation Economics, Engineering and Management Handbook 1: Main Themes
Innovation Economics, Engineering and Management Handbook 2: Special Themes

2020

ANDREOSSO-O'CALLAGHAN Bernadette, DZEVER Sam, JAUSSAUD Jacques, TAYLOR Robert
Sustainable Development and Energy Transition in Europe and Asia (Innovation and Technology Set – Volume 9)

BEN SLIMANE Sonia, M'HENNI Hatem
Entrepreneurship and Development: Realities and Future Prospects (Smart Innovation Set – Volume 30)

CHOUTEAU Marianne, FOREST Joëlle, NGUYEN Céline
Innovation for Society: The P.S.I. Approach (Smart Innovation Set – Volume 28)

CORON Clotilde
Quantifying Human Resources: Uses and Analysis
(Technological Changes and Human Resources Set – Volume 2)

CORON Clotilde, GILBERT Patrick
Technological Change
(Technological Changes and Human Resources Set – Volume 1)

CERDIN Jean-Luc, PERETTI Jean-Marie
The Success of Apprenticeships: Views of Stakeholders on Training and Learning (Human Resources Management Set – Volume 3)

DELCHET-COCHET Karen
Circular Economy: From Waste Reduction to Value Creation
(Economic Growth Set – Volume 2)

DIDAY Edwin, GUAN Rong, SAPORTA Gilbert, WANG Huiwen
Advances in Data Science
(Big Data, Artificial Intelligence and Data Analysis Set – Volume 4)

DOS SANTOS PAULINO Victor
Innovation Trends in the Space Industry
(Smart Innovation Set – Volume 25)

GUILHON Bernard
Venture Capital and the Financing of Innovation
(Innovation Between Risk and Reward Set – Volume 6)

JURCZENKO Emmanuel
Machine Learning for Asset Management: New Developments and Financial Applications

LATOUCHE Pascal
Open Innovation: Human Set-up
(Innovation and Technology Set – Volume 10)

MAKRIDES Andreas, KARAGRIGORIOU Alex, SKIADAS Christos H.
Data Analysis and Applications 3: Computational, Classification, Financial, Statistical and Stochastic Methods
(Big Data, Artificial Intelligence and Data Analysis Set – Volume 5)
Data Analysis and Applications 4: Financial Data Analysis and Methods
(Big Data, Artificial Intelligence and Data Analysis Set – Volume 6)

MASSOTTE Pierre, CORSI Patrick
Complex Decision-Making in Economy and Finance

MICHAUD Thomas
Science Fiction and Innovation Design
(Innovation in Engineering and Technology Set – Volume 6)

MONINO Jean-Louis
Data Control: Major Challenge for the Digital Society
(Smart Innovation Set – Volume 29)

MORLAT Clément
Sustainable Productive System: Eco-development versus Sustainable Development (Smart Innovation Set – Volume 26)

SAULAIS Pierre, ERMINE Jean-Louis
Knowledge Management in Innovative Companies 2: Understanding and Deploying a KM Plan within a Learning Organization
(Smart Innovation Set – Volume 27)

UZUNIDIS Dimitri
Systemic Innovation: Entrepreneurial Strategies and Market Dynamics

2019

AMENDOLA Mario, GAFFARD Jean-Luc
Disorder and Public Concern Around Globalization

BARBAROUX Pierre
Disruptive Technology and Defence Innovation Ecosystems
(Innovation in Engineering and Technology Set – Volume 5)

DOU Henri, JUILLET Alain, CLERC Philippe
Strategic Intelligence for the Future 1: A New Strategic and Operational Approach
Strategic Intelligence for the Future 2: A New Information Function Approach

FRIKHA Azza
Measurement in Marketing: Operationalization of Latent Constructs

FRIMOUSSE Soufyane
Innovation and Agility in the Digital Age
(Human Resources Management Set – Volume 2)

GAY Claudine, SZOSTAK Bérangère L.
Innovation and Creativity in SMEs: Challenges, Evolutions and Prospects
(Smart Innovation Set – Volume 21)

GORIA Stéphane, HUMBERT Pierre, ROUSSEL Benoît
Information, Knowledge and Agile Creativity
(Smart Innovation Set – Volume 22)

HELLER David
Investment Decision-making Using Optional Models
(Economic Growth Set – Volume 2)

HELLER David, DE CHADIRAC Sylvain, HALAOUI Lana, JOUVET Camille
The Emergence of Start-ups
(Economic Growth Set – Volume 1)

HÉRAUD Jean-Alain, KERR Fiona, BURGER-HELMCHEN Thierry
Creative Management of Complex Systems
(Smart Innovation Set – Volume 19)

LATOUCHE Pascal
Open Innovation: Corporate Incubator
(Innovation and Technology Set – Volume 7)

LEHMANN Paul-Jacques
The Future of the Euro Currency

LEIGNEL Jean-Louis, MÉNAGER Emmanuel, YABLONSKY Serge
Sustainable Enterprise Performance: A Comprehensive Evaluation Method

LIÈVRE Pascal, AUBRY Monique, GAREL Gilles
Management of Extreme Situations: From Polar Expeditions to Exploration-Oriented Organizations

MILLOT Michel
Embarrassment of Product Choices 2: Towards a Society of Well-being

N'GOALA Gilles, PEZ-PÉRARD Virginie, PRIM-ALLAZ Isabelle
Augmented Customer Strategy: CRM in the Digital Age

NIKOLOVA Blagovesta
The RRI Challenge: Responsibilization in a State of Tension with Market Regulation
(Innovation and Responsibility Set – Volume 3)

PELLEGRIN-BOUCHER Estelle, ROY Pierre
Innovation in the Cultural and Creative Industries
(Innovation and Technology Set – Volume 8)

PRIOLON Joël
Financial Markets for Commodities

QUINIOU Matthieu
Blockchain: The Advent of Disintermediation

RAVIX Joël-Thomas, DESCHAMPS Marc
Innovation and Industrial Policies
(Innovation between Risk and Reward Set – Volume 5)

ROGER Alain, VINOT Didier
Skills Management: New Applications, New Questions
(Human Resources Management Set – Volume 1)

SAULAIS Pierre, ERMINE Jean-Louis
Knowledge Management in Innovative Companies 1: Understanding and Deploying a KM Plan within a Learning Organization
(Smart Innovation Set – Volume 23)

SERVAJEAN-HILST Romaric
Co-innovation Dynamics: The Management of Client-Supplier Interactions for Open Innovation
(Smart Innovation Set – Volume 20)

SKIADAS Christos H., BOZEMAN James R.
Data Analysis and Applications 1: Clustering and Regression, Modeling-estimating, Forecasting and Data Mining
(Big Data, Artificial Intelligence and Data Analysis Set – Volume 2)
Data Analysis and Applications 2: Utilization of Results in Europe and Other Topics
(Big Data, Artificial Intelligence and Data Analysis Set – Volume 3)

VIGEZZI Michel
World Industrialization: Shared Inventions, Competitive Innovations and Social Dynamics
(Smart Innovation Set – Volume 24)

2018

BURKHARDT Kirsten
Private Equity Firms: Their Role in the Formation of Strategic Alliances

CALLENS Stéphane
Creative Globalization
(Smart Innovation Set – Volume 16)

CASADELLA Vanessa
Innovation Systems in Emerging Economies: MINT – Mexico, Indonesia, Nigeria, Turkey
(Smart Innovation Set – Volume 18)

CHOUTEAU Marianne, FOREST Joëlle, NGUYEN Céline
Science, Technology and Innovation Culture
(Innovation in Engineering and Technology Set – Volume 3)

CORLOSQUET-HABART Marine, JANSSEN Jacques
Big Data for Insurance Companies
(Big Data, Artificial Intelligence and Data Analysis Set – Volume 1)

CROS Françoise
Innovation and Society
(Smart Innovation Set – Volume 15)

DEBREF Romain
Environmental Innovation and Ecodesign: Certainties and Controversies
(Smart Innovation Set – Volume 17)

DOMINGUEZ Noémie
SME Internationalization Strategies: Innovation to Conquer New Markets

ERMINE Jean-Louis
Knowledge Management: The Creative Loop
(Innovation and Technology Set – Volume 5)

GAUDIN Thierry, LACROIX Dominique, MAUREL Marie-Christine, POMEROL, Jean-Charles
Life Sciences, Information Sciences

GILBERT Patrick, BOBADILLA Natalia, GASTALDI Lise, LE BOULAIRE Martine, LELEBINA Olga
Innovation, Research and Development Management

VAN HOOREBEKE Delphine
The Management of Living Beings or Emo-management

IBRAHIMI Mohammed
Mergers & Acquisitions: Theory, Strategy, Finance

LEMAÎTRE Denis
Training Engineers for Innovation

LÉVY Aldo, BEN BOUHENI Faten, AMMI Chantal
Financial Management: USGAAP and IFRS Standards
(Innovation and Technology Set – Volume 6)

LOISY Catherine, COULET Jean-Claude
Competence and Program-based Approach in Training: Tools for Developing Responsible Activities
Innovation and Responsibility Set – Volume 1)

MILLOT Michel
Embarrassment of Product Choices 1: How to Consume Differently

PANSERA Mario, OWEN Richard
Innovation and Development: The Politics at the Bottom of the Pyramid (Innovation and Responsibility Set – Volume 2)

RICHEZ Yves
Corporate Talent Detection and Development

SACHETTI Philippe, ZUPPINGER Thibaud
New Technologies and Branding (Innovation and Technology Set – Volume 4)

SAMIER Henri
Intuition, Creativity, Innovation

TEMPLE Ludovic, COMPAORÉ SAWADOGO Eveline M.F.W.
Innovation Processes in Agro-Ecological Transitions in Developing Countries (Innovation in Engineering and Technology Set – Volume 2)

UZUNIDIS Dimitri
Collective Innovation Processes: Principles and Practices (Innovation in Engineering and Technology Set – Volume 4)

2017

AÏT-EL-HADJ Smaïl
The Ongoing Technological System (Smart Innovation Set – Volume 11)

D'ANDRIA Aude, GABARRET Inés
Building 21st Century Entrepreneurship (Innovation and Technology Set – Volume 2)

BAUDRY Marc, DUMONT Béatrice
Patents: Prompting or Restricting Innovation? (Smart Innovation Set – Volume 12)

BÉRARD Céline, TEYSSIER Christine
Risk Management: Lever for SME Development and Stakeholder Value Creation

CHALENÇON Ludivine
Location Strategies and Value Creation of International Mergers and Acquisitions

CHAUVEL Danièle, BORZILLO Stefano
The Innovative Company: An Ill-defined Object
(Innovation between Risk and Reward Set – Volume 1)

CORSI Patrick
Going Past Limits To Growth

DAIDJ Nabyla
Cooperation, Coopetition and Innovation
(Innovation and Technology Set – Volume 3)

FERNEZ-WALCH Sandrine
The Multiple Facets of Innovation Project Management
(Innovation between Risk and Reward Set – Volume 4)

FOREST Joëlle
Creative Rationality and Innovation
(Smart Innovation Set – Volume 14)

GUILHON Bernard
Innovation and Production Ecosystems
(Innovation between Risk and Reward Set – Volume 2)

HAMMOUDI Abdelhakim, DAIDJ Nabyla
Game Theory Approach to Managerial Strategies and Value Creation
(Diverse and Global Perspectives on Value Creation Set – Volume 3)

LALLEMENT Rémi
Intellectual Property and Innovation Protection: New Practices and New Policy Issues
(Innovation between Risk and Reward Set – Volume 3)

LAPERCHE Blandine
Enterprise Knowledge Capital
(Smart Innovation Set – Volume 13)

LEBERT Didier, EL YOUNSI Hafida
International Specialization Dynamics
(Smart Innovation Set – Volume 9)

MAESSCHALCK Marc
Reflexive Governance for Research and Innovative Knowledge
(Responsible Research and Innovation Set – Volume 6)

MASSOTTE Pierre
Ethics in Social Networking and Business 1: Theory, Practice and Current Recommendations
Ethics in Social Networking and Business 2: The Future and Changing Paradigms

MASSOTTE Pierre, CORSI Patrick
Smart Decisions in Complex Systems

MEDINA Mercedes, HERRERO Mónica, URGELLÉS Alicia
Current and Emerging Issues in the Audiovisual Industry
(Diverse and Global Perspectives on Value Creation Set – Volume 1)

MICHAUD Thomas
Innovation, Between Science and Science Fiction
(Smart Innovation Set – Volume 10)

PELLÉ Sophie
Business, Innovation and Responsibility
(Responsible Research and Innovation Set – Volume 7)

SAVIGNAC Emmanuelle
The Gamification of Work: The Use of Games in the Workplace

SUGAHARA Satoshi, DAIDJ Nabyla, USHIO Sumitaka
Value Creation in Management Accounting and Strategic Management: An Integrated Approach
(Diverse and Global Perspectives on Value Creation Set –Volume 2)

UZUNIDIS Dimitri, SAULAIS Pierre
Innovation Engines: Entrepreneurs and Enterprises in a Turbulent World (Innovation in Engineering and Technology Set – Volume 1)

2016

BARBAROUX Pierre, ATTOUR Amel, SCHENK Eric
Knowledge Management and Innovation (Smart Innovation Set – Volume 6)

BEN BOUHENI Faten, AMMI Chantal, LEVY Aldo
Banking Governance, Performance And Risk-Taking: Conventional Banks Vs Islamic Banks

BOUTILLIER Sophie, CARRÉ Denis, LEVRATTO Nadine
Entrepreneurial Ecosystems (Smart Innovation Set – Volume 2)

BOUTILLIER Sophie, UZUNIDIS Dimitri
The Entrepreneur (Smart Innovation Set – Volume 8)

BOUVARD Patricia, SUZANNE Hervé
Collective Intelligence Development in Business

GALLAUD Delphine, LAPERCHE Blandine
Circular Economy, Industrial Ecology and Short Supply Chains (Smart Innovation Set – Volume 4)

GUERRIER Claudine
Security and Privacy in the Digital Era (Innovation and Technology Set – Volume 1)

MEGHOUAR Hicham
Corporate Takeover Targets

MONINO Jean-Louis, SEDKAOUI Soraya
Big Data, Open Data and Data Development (Smart Innovation Set – Volume 3)

MOREL Laure, LE ROUX Serge
Fab Labs: Innovative User
(Smart Innovation Set – Volume 5)

PICARD Fabienne, TANGUY Corinne
Innovations and Techno-ecological Transition
(Smart Innovation Set – Volume 7)

2015

CASADELLA Vanessa, LIU Zeting, DIMITRI Uzunidis
Innovation Capabilities and Economic Development in Open Economies
(Smart Innovation Set – Volume 1)

CORLOSQUET-HABART Marine, GEHIN William, JANSSEN Jaques, MANCA Raimondo
Asset and Liabilities Management for Banks and Insurance Companies

CORSI Patrick, MORIN Dominique
Sequencing Apple's DNA

CORSI Patrick, NEAU Erwan
Innovation Capability Maturity Model

FAIVRE-TAVIGNOT Bénédicte
Social Business and Base of the Pyramid

GODÉ Cécile
Team Coordination in Extreme Environments

MAILLARD Pierre
Competitive Quality and Innovation

MASSOTTE Pierre, CORSI Patrick
Operationalizing Sustainability

MASSOTTE Pierre, CORSI Patrick
Sustainability Calling

2014

DUBÉ Jean, LEGROS Diègo
Spatial Econometrics Using Microdata

LESCA Humbert, LESCA Nicolas
Strategic Decisions and Weak Signals

2013

HABART-CORLOSQUET Marine, JANSSEN Jacques, MANCA Raimondo
VaR Methodology for Non-Gaussian Finance

2012

POMEROL Jean-Charles
Decision-Making and Action

SZYLAR Christian
UCITS Handbook

2011

LESCA Nicolas
Environmental Scanning and Sustainable Development

LESCA Nicolas, LESCA Humbert
Weak Signals for Strategic Intelligence: Anticipation Tool for Managers

MERCIER-LAURENT Eunika
Innovation Ecosystems

2010

SZYLAR Christian
Risk Management under UCITS III/IV

2009

COHEN Corine
Business Intelligence

ZANINETTI Jean-Marc
Sustainable Development in the USA

2008

CORSI Patrick, DULIEU Mike
The Marketing of Technology Intensive Products and Services

DZEVER Sam, JAUSSAUD Jacques, ANDREOSSO Bernadette
Evolving Corporate Structures and Cultures in Asia: Impact of Globalization

2007

AMMI Chantal
Global Consumer Behavior

2006

BOUGHZALA Imed, ERMINE Jean-Louis
Trends in Enterprise Knowledge Management

CORSI Patrick, CHRISTOFOL Hervé, RICHIR Simon, SAMIER Henri
Innovation Engineering: the power of intangible networks

Printed and bound by CPI Group (UK) Ltd, Croydon, CR0 4YY

13/09/2026

14961013-0006